AF134269

AFRICAN DIASPORA MATHEMATICS RESEARCH PROGRESS

AFRICAN DIASPORA MATHEMATICS RESEARCH PROGRESS

TOKA DIAGANA
EDITOR

Nova Science Publishers, Inc.
New York

Copyright © 2008 by Nova Science Publishers, Inc.

All rights reserved. No part of this book may be reproduced, stored in a retrieval system or transmitted in any form or by any means: electronic, electrostatic, magnetic, tape, mechanical photocopying, recording or otherwise without the written permission of the Publisher.

For permission to use material from this book please contact us:
Telephone 631-231-7269; Fax 631-231-8175
Web Site: http://www.novapublishers.com

NOTICE TO THE READER

The Publisher has taken reasonable care in the preparation of this book, but makes no expressed or implied warranty of any kind and assumes no responsibility for any errors or omissions. No liability is assumed for incidental or consequential damages in connection with or arising out of information contained in this book. The Publisher shall not be liable for any special, consequential, or exemplary damages resulting, in whole or in part, from the readers' use of, or reliance upon, this material.

Independent verification should be sought for any data, advice or recommendations contained in this book. In addition, no responsibility is assumed by the publisher for any injury and/or damage to persons or property arising from any methods, products, instructions, ideas or otherwise contained in this publication.

This publication is designed to provide accurate and authoritative information with regard to the subject matter cover herein. It is sold with the clear understanding that the Publisher is not engaged in rendering legal or any other professional services. If legal, medical or any other expert assistance is required, the services of a competent person should be sought. FROM A DECLARATION OF PARTICIPANTS JOINTLY ADOPTED BY A COMMITTEE OF THE AMERICAN BAR ASSOCIATION AND A COMMITTEE OF PUBLISHERS.

Library of Congress Cataloging-in-Publication Data

African diaspora mathematics research progress / Toka Diagana, editor.
p. cm.
ISBN 978-1-60456-204-0 (hardcover)
1. Mathematics--Research--Africa. 2. Mathematics--Study and teaching--Africa. 3. African diaspora. I. Diagana, Toka.
QA14.A35A37 2008
510--dc22 2008011668

Published by Nova Science Publishers, Inc. ✦ New York

CONTENTS

PREFACE

This book presents carefully selected mathematical contributions from the African Diaspora. The book is based on mathematical research with some emphasis on the contributions of all African mathematicians and the rich connections between all African universities and those of other continents. This includes the Weighted Boundedness for Multilinear Singular Integral Operator, Measure Theory, Pseudo-Differential Operators and Commutators in Multiplier Spaces, RBSDEs with Stochastic Monotone, Collapse of a Void Spherical Bubble Immersed in a Non-Newtonian Fluid, Hazard Rate Prediction in Life Time Data Analysis, Singular Reduction and Stratification of Quiver Variety, and Mathieu Function and Kontorovich-Lebedev Transforms.

Chapter 1 - This paper proves the weighted boundedness for some multilinear singular integral operators with variable Calderón-Zygmund kernel on L^p and Morrey spaces.

Chapter 2 – The authors construct an integral relatively to a closed convex-valued multilinear measure. This method unifies all the different integrals constructed by both D. S. Thiam and Pallu De La Barrière. For that, the authors introduce the notion of s-compact, which is for set-valued measures, what σ-finite is for scalar measures. The authors introduce also a notion of negligible functions. The authors apply these notions to the construction of the integral. The authors introduce a topology of the convergence in mean, for which, the spaces of integrable functions are complete.

Chapter 3 - In this paper the authors establish the boundedness of pseudo-differential operators with symbols in the class $S^{(-n\theta/2)}{}_{1-\alpha,\delta}$, $(0 < \theta < 1, \delta < 1 - \theta)$ and their commutators with *BMO* functions in multiplier spaces. As a consequence of this result, the authors extend some results on L^p by extrapolation.

Chapter 4 - In this paper, the authors are concerned with reflected backward stochastic differential equations (RBSDEs) in a domain of a lower semi-continuous convex function with stochastic monotone and polynomial growth generators. The authors prove existence and uniqueness result for fixed terminal time. This work provides an extension of the result established under uniform monotonicity condition.

Chapter 5 - In this paper the authors establish a pinching condition to insure that submanifolds of codimension $p \geq 2$ in the unit sphere are spheres.

Chapter 6 - The authors study analytically the dynamics of a single void spherical bubble immersed in a power-law non-Newtonian fluid and in a second-grade fluid. The authors derive the equation of motion of the bubble wall and the authors prove that it is integrable. The authors establish that near collapse, the radius of the bubble behaves like $(t_c\text{-}t)^k$, where k

$\in \{2/5, (2 - n)/2\}$ for a power-law fluid of index n, $k \in \{1/2, 1/3\}$ for a second-grade fluid, and t_c is the collapse time.

Chapter 7 – The authors consider in this paper a nonparametric estimation of the hazard rate function based on right-censored data using the wavelets method. Asymptotic properties and strong uniform consistency rates are established under suitable conditions.

Chapter 8 - In this article the authors study singular reduction and stratification in the case of the action of a complex reductive Lie group on a Quiver Variety. The main result of the paper is an illustration of the key role R. Sjamaar's Holomorphic Slice Theorem can play in the understanding of some interesting aspects of singular reduction theory.

Chapter 9 - The authors consider a boundary-value problem for the Helmholtz equation outside a right-angled wedge configuration formed by a half-plane and a strip (i.e., the so-called L-shaped surface boundary). The problem models the diffraction of plane waves by scatterers of such L-shaped configurations. The proposed scheme for the solution of the problem includes an application of the Kontorovich-Lebedev (KL) transform and a new discrete index of the Mathieu function (diMf) transform. Within the present approach, an integral equation satisfied by the KL spectrum, and a linear system for the diMf spectral amplitudes are derived. In addition, the singularities of the spectral function are deduced. Moreover, near and far field representations are also obtained.

In: African Diaspora Mathematics Research Progress
Editor: Toka Diagana, pp. 1–12
ISBN 978-1-60456-204-0
© 2008 Nova Science Publishers, Inc.

Chapter 1

WEIGHTED BOUNDEDNESS FOR MULTILINEAR SINGULAR INTEGRAL OPERATOR WITH VARIABLE CALDERÓN-ZYGMUND KERNEL

Hong Xu and Lanzhe Liu*
Department of Mathematics, Changsha University of
Science and Technology, Changsha 410077, P.R.of China
Yueyang Vocational Technical College
Yueyang 414000, P. R. of China

Abstract

In this paper, we prove the weighted boundedness for some multilinear singular integral operators with variable Calderón-Zygmund kernel on L^p and Morrey spaces.

AMS Subject Classification: 42B20, 42B25.

Keywords: Multilinear singular integral operator; variable Calderón-Zygmund kernel; BMO; Morrey spaces.

1 Introduction

As the development of Calderón-Zygmund operators and their commutators, singular integral operators have been well studied. In [1], Calderón and Zygmund introduce some singular integral operators with variable kernel and discuss their boundedness. In [9], G.Di Fazio and M. A. Ragusa obtain the boundedness for the commutators generated by the singular integral operators with variable kernel and BMO functions. In [11], Lu and al. prove the boundedness for the multilinear oscillatory singular integral operators generated by the operators and BMO functions. In this paper, we will study the multilinear singular integral operators with variable Calderón-Zygmund kernel.

*E-mail address:lanzheliu@163.com

Definition 1.1. Let $k(x) = \Omega(x)/|x|^n : R^n \setminus \{0\} \longrightarrow R$. k is said to be a Calderón-Zygmund kernel if

(a) $\Omega \in C^\infty(R^n \setminus \{0\})$,

(b) Ω is homogeneous of degree zero,

(c) $\int_\Sigma \Omega(x) x^\alpha d\sigma(x) = 0$ for all multi-indices $\alpha \in (N \bigcup \{0\})^n$ with $|\alpha| = n$,

where $\Sigma = \{x \in R^n : |x| = 1\}$ is the unit sphere of R^n.

Definition 1.2. Let $k(x,y) = \Omega(x,y)/|y|^n : R^n \times (R^n \setminus \{0\}) \longrightarrow R$. k is said to be a variable Calderón-Zygmund kernel if

(d) $k(x,\cdot)$ is a Calderón-Zygmund kernel for a.e. $x \in R^n$;

(e) $\max_{|\gamma| \le 2n} \left|\left| \frac{\partial^{|\gamma|}}{\partial^\gamma y} \Omega(x,y) \right|\right|_{L^\infty(R^n \times \Sigma)} = M < \infty$.

Let m_j be the positive integers($j = 1, \cdots, l$), $m_1 + \cdots + m_l = m$ and A_j be the functions on R^n ($j = 1, \cdots, l$). Denote

$$R_{m_j+1}(A_j;x,y) = A_j(x) - \sum_{|\alpha| \le m_j} \frac{1}{\alpha!} D^\alpha A_j(y)(x-y)^\alpha.$$

The multilinear singular integral operator with variable Calderón-Zygmund kernel is defined by

$$T^A(f)(x) = \int_{R^n} \frac{\Omega(x, x-y)}{|x-y|^{n+m}} \prod_{j=1}^{l} R_{m_j+1}(A_j;x,y) f(y) dy,$$

where $\Omega(x,y)/|y|^n$ is a variable Calderón-Zygmund kernel. We also define

$$T(f)(x) = \int_{R^n} \frac{\Omega(x,x-y)}{|x-y|^n} f(y) dy,$$

which is the singular integral operator with variable Calderón-Zygmund kernel(see [1]).

Note that when $m = 0$, T^A is just the multilinear commutator of T and A(see [11]). While when $m > 0$, T^A is the non-trivial generalizations of the commutator. It is well known that multilinear operators are of great interest in harmonic analysis and have been widely studied by many authors (see [3, 4, 5, 6, 7]). Cohen and Gosselin(see [3, 4, 5]) obtained the $L^p(p > 1)$ boundedness of the multilinear singular integral operator. The main purpose of this paper is to prove the weighted L^p and Morrey spaces boundedness for the multilinear singular integral operators with variable Calderón-Zygmund kernel.

First, let us introduce some notations. Throughout this paper, Q will denote a cube of R^n with sides parallel to the axes. For a set E and a non-negative locally integrable function w, let $w(E) = \int_E w(x)dx$. For any locally integrable function f, the sharp function of f is defined by

$$f^\#(x) = \sup_{x \in Q} \frac{1}{|Q|} \int_Q |f(y) - f_Q| dy,$$

where, and in what follows, $f_Q = |Q|^{-1} \int_Q f(x)dx$. It is well-known that ([10])

$$f^\#(x) = \sup_{x \in Q} \inf_{c \in C} \frac{1}{|Q|} \int_Q |f(y) - c| dy.$$

We say that f belongs to $BMO(R^n)$ if $f^{\#}$ belongs to $L^{\infty}(R^n)$ and $||f||_{BMO} = ||f^{\#}||_{L^{\infty}}$. Let M be the Hardy-Littlewood maximal operator defined by

$$M(f)(x) = \sup_{x\in Q} |Q|^{-1} \int_Q |f(y)|dy,$$

we write that $M_p(f) = (M(f^p))^{1/p}$ for $0 < p < \infty$. Let A_1 be the Muckenhoupt's class(see e.g. [10]):

$$A_1 = \{0 < w \in L^1_{loc}(R^n) : M(w)(x) \le Cw(x), a.e.\}.$$

Let φ be a positive, increasing function on R^+ and there exists a constant $D > 0$ such that

$$\varphi(2t) \le D\varphi(t) \ \text{ for } \ t \ge 0.$$

Let w be a non-negative weight function on R^n and f be a locally integrable function on R^n. Define that, for $1 \le p < \infty$,

$$||f||_{L^{p,\varphi}(w)} = \sup_{x\in R^n,\ d>0} \left(\frac{1}{\varphi(d)} \int_{B(x,d)} |f(y)|^p w(y)dy\right)^{1/p},$$

where $B(x,d) = \{y \in R^n : |x-y| < d\}$. The generalized weighted Morrey spaces is defined by

$$L^{p,\varphi}(w) = \{f \in L^1_{loc}(R^n) : ||f||_{L^{p,\varphi}(w)} < \infty\}.$$

If $\varphi(d) = d^{\delta}$, $\delta > 0$, then $L^{p,\varphi}(w) = L^{p,\delta}(w)$, which is the classical Morrey spaces (see [12]).

We shall prove the following theorems.

Theorem 1.3. *Let $1 < p < \infty$, $w \in A_1$ and $D^{\alpha}A_j \in BMO(R^n)$ for all α with $|\alpha| = m_j$ and $j = 1,\cdots,l$. Then T^A is bounded on $L^p(w)$, that is*

$$||T^A(f)||_{L^p(w)} \le C\prod_{j=1}^{l}\left(\sum_{|\alpha_j|=m_j} ||D^{\alpha_j}A_j||_{BMO}\right)||f||_{L^p(w)}.$$

Theorem 1.4. *Let $1 < p < \infty$, $0 < D < 2^n$, $w \in A_1$ and $D^{\alpha}A_j \in BMO(R^n)$ for all α with $|\alpha| = m_j$ and $j = 1,\cdots,l$. Then T^A is bounded on $L^{p,\varphi}(w)$, that is*

$$||T^A(f)||_{L^{p,\varphi}(w)} \le C\prod_{j=1}^{l}\left(\sum_{|\alpha_j|=m_j} ||D^{\alpha_j}A_j||_{BMO}\right)||f||_{L^{p,\varphi}(w)}.$$

2 Proof of Theorems

To prove the theorems, we need the following lemmas.

Lemma 2.1. *([5]) Let A be a function on R^n and $D^{\alpha}A \in L^q(R^n)$ for all α with $|\alpha| = m$ and some $q > n$. Then*

$$|R_m(A;x,y)| \le C|x-y|^m \sum_{|\alpha|=m}\left(\frac{1}{|\tilde{Q}(x,y)|}\int_{\tilde{Q}(x,y)} |D^{\alpha}A(z)|^q dz\right)^{1/q},$$

where $\tilde{Q}$ is the cube centered at x and having side length $5\sqrt{n}|x-y|$.

Lemma 2.2. *Let B be a ball of R^n and $w \in \Lambda_1$. Then $M(w\chi_B) \in \Lambda_1$.*

Proof. We need only prove that, for any cube Q,

$$\frac{1}{|Q|}\int_Q M(w\chi_B)(y)dy \le ess \inf_{x\in Q} M(w\chi_B)(x).$$

We first prove that, for any $x, z \in Q$,

$$M(w\chi_B)(x) \le CM(w\chi_B)(z).$$

In fact, let Q' be any cube with $Q' \ni x$. Consider

$$\frac{1}{|Q'|}\int_{Q'} w(y)\chi_B(y)dy.$$

Without loss of generality, we may assume $Q' \bigcap B \neq \emptyset$. Then, there exist $c > 0$ such that $z \in cQ'$, thus

$$\frac{1}{|Q'|}\int_{Q'} w(y)\chi_B(y)dy \le CM(w\chi_B)(z)$$

and

$$M(w\chi_B)(x) = \sup_{x\in Q'} \frac{1}{|Q'|}\int_{Q'} w(y)\chi_B(y)dy \le CM(w\chi_B)(z).$$

Since z and x are arbitrary, we conclude

$$M(w\chi_B)(x) \le C \inf_{z\in Q} M(w\chi_B)(z)$$

and

$$\frac{1}{|Q|}\int_Q M(w\chi_B)(x)dx \le C \inf_{z\in Q} M(w\chi_B)(z),$$

which is what we wanted.

Proof of Theorem 1.3. It suffices to prove the theorem for $f \in C_0^\infty(R^n)$. Without loss of generality, we may assume $l = 2$. Fix a cube $Q = Q(x_0, d)$ and $\tilde{x} \in Q$. Let $\tilde{Q} = 5\sqrt{n}Q$ and $\tilde{A}_j(x) = A_j(x) - \sum_{|\alpha|=m_j} \frac{1}{\alpha!}(D^\alpha A_j)_{\tilde{Q}} x^\alpha$, then $R_{m_j}(A_j;x,y) = R_{m_j}(\tilde{A}_j;x,y)$ and $D^\alpha \tilde{A}_j = D^\alpha A_j - (D^\alpha A_j)_{\tilde{Q}}$ for $|\alpha| = m_j$. We write, for $f_1 = f\chi_{\tilde{Q}}$ and $f_2 = f\chi_{R^n\setminus\tilde{Q}}$,

$$|T^A(f)(x)| \le \left|\int_{R^n} \frac{\Omega(x,x-y)}{|x-y|^{n+m}} \prod_{j=1}^{2} R_{m_j}(\tilde{A}_j;x,y) f_1(y)dy\right| \tag{2.1}$$

$$+C\left|\sum_{|\alpha_1|=m_1} \int_{R^n} \frac{\Omega(x,x-y)}{|x-y|^{n+m}} R_{m_2}(\tilde{A}_2;x,y)(x-y)^{\alpha_1} D^{\alpha_1}\tilde{A}_1(y) f_1(y)dy\right| \tag{2.2}$$

$$+C\left|\sum_{|\alpha_2|=m_2} \int_{R^n} \frac{\Omega(x,x-y)}{|x-y|^{n+m}} R_{m_1}(\tilde{A}_1;x,y)(x-y)^{\alpha_2} D^{\alpha_2}\tilde{A}_2(y) f_1(y)dy\right| \tag{2.3}$$

$$+C\left|\sum_{|\alpha_1|=m_1,|\alpha_2|=m_2} \int_{R^n} \frac{\Omega(x,x-y)}{|x-y|^{n+m}} (x-y)^{\alpha_1+\alpha_2} D^{\alpha_1}\tilde{A}_1(y) D^{\alpha_2}\tilde{A}_2(y) f_1(y)dy\right| \tag{2.4}$$

$$+|T^{\tilde{A}}(f_2)(x)| \tag{2.5}$$

$$:= I_1(x) + I_2(x) + I_3(x) + I_4(x) + I_5(x). \tag{2.6}$$

Now, let us estimate $I_1(x)$, $I_2(x)$, $I_3(x)$, $I_4(x)$ and $I_5(x)$, respectively. Fixed $1 < q < p$. First, for $x \in Q$ and $y \in \tilde{Q}$, by Lemma 2.1, we get

$$R_{m_j}(\tilde{A}_j;x,y) \le C|x-y|^{m_j} \sum_{|\alpha_j|=m_j} ||D^{\alpha_j}A_j||_{BMO},$$

by the L^q-boundedness of T, we obtain

$$\begin{aligned}
&\frac{1}{|Q|}\int_Q I_1(x)dx && (2.7)\\
&\le C\prod_{j=1}^{2}\left(\sum_{|\alpha_j|=m_j} ||D^{\alpha_j}A_j||_{BMO}\right)\frac{1}{|Q|}\int_Q |T(f_1)(x)|dx && (2.8)\\
&\le C\prod_{j=1}^{2}\left(\sum_{|\alpha_j|=m_j} ||D^{\alpha_j}A_j||_{BMO}\right)\left(\frac{1}{|Q|}\int_{R^n} |T(f_1)(x)|^q dx\right)^{1/q} && (2.9)\\
&\le C\prod_{j=1}^{2}\left(\sum_{|\alpha_j|=m_j} ||D^{\alpha_j}A_j||_{BMO}\right)|Q|^{-1/q}\left(\int_{R^n} |f_1(x)|^q dx\right)^{1/q} && (2.10)\\
&\le C\prod_{j=1}^{2}\left(\sum_{|\alpha_j|=m_j} ||D^{\alpha_j}A_j||_{BMO}\right)M_q(f)(\tilde{x}), && (2.11)
\end{aligned}$$

thus, $M(I_1)(\tilde{x}) \le C\prod_{j=1}^{2}\left(\sum_{|\alpha|=m_j} ||D^{\alpha}A_j||_{BMO}\right)M_q(f)(\tilde{x})$ and

$$\begin{aligned}
||I_1(\cdot)||_{L^p(w)} &\le ||M(I_1)||_{L^p(w)} && (2.12)\\
&\le C\prod_{j=1}^{2}\left(\sum_{|\alpha_j|=m_j} ||D^{\alpha_j}A_j||_{BMO}\right)||M_q(f)||_{L^p(w)} && (2.13)\\
&\le C\prod_{j=1}^{2}\left(\sum_{|\alpha_j|=m_j} ||D^{\alpha_j}A_j||_{BMO}\right)||f||_{L^p(w)}. && (2.14)
\end{aligned}$$

For $I_2(x)$, taking $q = rs$ for $1 < r,s < \infty$ and $1/s + 1/s' = 1$, we have, by the L^r-

boundedness of T and Hölder's inequality,

$$\frac{1}{|Q|}\int_Q I_2(x)dx \le C\sum_{|\alpha_2|=m_2}||D^{\alpha_2}A_2||_{BMO}\sum_{|\alpha_1|=m_1}\frac{1}{|Q|}\int_Q |T(D^{\alpha_1}\tilde{A}_1 f_1)(x)|dx \quad (2.15)$$

$$\le C\sum_{|\alpha_2|=m_2}||D^{\alpha_2}A_2||_{BMO}\left(\sum_{|\alpha_1|=m_1}\frac{1}{|Q|}\int_{R^n}|T(D^{\alpha_1}\tilde{A}_1 f_1)(x)|^r dx\right)^{1/r} \quad (2.16)$$

$$\le C\sum_{|\alpha_2|=m_2}||D^{\alpha_2}A_2||_{BMO}\sum_{|\alpha_1|=m_1}\left(\frac{1}{|Q|}\int_{R^n}|D^{\alpha_1}\tilde{A}_1(x)f_1(x)|^r dx\right)^{1/r} \quad (2.17)$$

$$\le C\sum_{|\alpha_2|=m_2}||D^{\alpha_2}A_2||_{BMO}\left(\frac{1}{|Q|}\int_{\tilde{Q}}|f(x)|^{rs}dx\right)^{1/rs} \quad (2.18)$$

$$\times\sum_{|\alpha_1|=m_1}\left(\frac{1}{|Q|}\int_{\tilde{Q}}|D^{\alpha_1}\tilde{A}_1(x)-(D^{\alpha}A_1)_{\tilde{Q}}|^{rs'}dx\right)^{1/rs'} \quad (2.19)$$

$$\le C\prod_{j=1}^{2}\left(\sum_{|\alpha|=m_j}||D^{\alpha}A_j||_{BMO}\right)M_q(f)(\tilde{x}), \quad (2.20)$$

thus, $M(I_2)(\tilde{x}) \le C\prod_{j=1}^{2}\left(\sum_{|\alpha|=m_j}||D^{\alpha}A_j||_{BMO}\right)M_q(f)(\tilde{x})$ and

$$||I_2(\cdot)||_{L^p(w)} \le ||M(I_2)||_{L^p(w)} \quad (2.21)$$

$$\le C\prod_{j=1}^{2}\left(\sum_{|\alpha_j|=m_j}||D^{\alpha_j}A_j||_{BMO}\right)||M_q(f)||_{L^p(w)} \quad (2.22)$$

$$\le C\prod_{j=1}^{2}\left(\sum_{|\alpha_j|=m_j}||D^{\alpha_j}A_j||_{BMO}\right)||f||_{L^p(w)}. \quad (2.23)$$

For $I_3(x)$, similar to the proof of $I_2(x)$, we get

$$||I_3(\cdot)||_{L^p(w)} \le C\prod_{j=1}^{2}\left(\sum_{|\alpha_j|=m_j}||D^{\alpha_j}A_j||_{BMO}\right)||f||_{L^p(w)}.$$

Similarly, for $I_4(x)$, denoting $q=rs_3$ for $1<r,s_1,s_2,s_3<\infty$ and $1/s_1+1/s_2+1/s_3=1$, we obtain

$$\frac{1}{|Q|}\int_Q I_4(x)dx \leq C\sum_{|\alpha_1|=m_1,|\alpha_2|=m_2}\frac{1}{|Q|}\int_Q |T(D^{\alpha_1}\tilde{A}_1 D^{\alpha_2}\tilde{A}_2 f_1)(x)|dx \quad (2.24)$$

$$\leq C\sum_{|\alpha_1|=m_1,|\alpha_2|=m_2}\left(\frac{1}{|Q|}\int_{R^n}|T(D^{\alpha_1}\tilde{A}_1 D^{\alpha_2}\tilde{A}_2 f_1)(x)|^r dx\right)^{1/r} \quad (2.25)$$

$$\leq C\sum_{|\alpha_1|=m_1,|\alpha_2|=m_2}|Q|^{-1/r}\left(\int_{R^n}|D^{\alpha_1}\tilde{A}_1(x)D^{\alpha_2}\tilde{A}_2(x)f_1(x)|^r dx\right)^{1/r} \quad (2.26)$$

$$\leq C\sum_{|\alpha_1|=m_1,|\alpha_2|=m_2}\left(\frac{1}{|Q|}\int_{\tilde{Q}}|D^{\alpha_1}\tilde{A}_1(x)|^{rs_1}dx\right)^{1/rs_1} \quad (2.27)$$

$$\times\left(\frac{1}{|Q|}\int_{\tilde{Q}}|D^{\alpha_2}\tilde{A}_2(x)|^{rs_2}dx\right)^{1/rs_2}\left(\frac{1}{|Q|}\int_{\tilde{Q}}|f(x)|^{rs_3}dx\right)^{1/rs_3} \quad (2.28)$$

$$\leq C\prod_{j=1}^{2}\left(\sum_{|\alpha|=m_j}||D^{\alpha}A_j||_{BMO}\right)M_q(f)(\tilde{x}), \quad (2.29)$$

thus, $M(I_4)(\tilde{x}) \leq C\prod_{j=1}^{2}\left(\sum_{|\alpha|=m_j}||D^{\alpha}A_j||_{BMO}\right)M_q(f)(\tilde{x})$ and

$$||I_4(\cdot)||_{L^p(w)} \leq ||M(I_4)||_{L^p(w)} \quad (2.30)$$

$$\leq C\prod_{j=1}^{2}\left(\sum_{|\alpha_j|=m_j}||D^{\alpha_j}A_j||_{BMO}\right)||M_q(f)||_{L^p(w)} \quad (2.31)$$

$$\leq C\prod_{j=1}^{2}\left(\sum_{|\alpha_j|=m_j}||D^{\alpha_j}A_j||_{BMO}\right)||f||_{L^p(w)}. \quad (2.32)$$

For $I_5(x)$, by [2], we know that

$$T^A(f)(x) = \sum_{k=1}^{\infty}\sum_{h=1}^{g_k}a_{hk}(x)\int_{R^n}\frac{Y_{hk}(x-y)}{|x-y|^{n+m}}\prod_{j=1}^{2}R_{m_j+1}(A_j;x,y)f(y)dy \quad (2.33)$$

$$= \sum_{k=1}^{\infty}\sum_{h=1}^{g_k}a_{hk}(x)S_{hk}^{A}(f)(x), \quad (2.34)$$

where $g_k \leq Ck^{n-2}$, $||a_{hk}||_{L^\infty} \leq Ck^{-2n}$, $|Y_{hk}(x-y)| \leq Ck^{n/2-1}$ and

$$\left|\frac{Y_{hk}(x-y)}{|x-y|^n} - \frac{Y_{hk}(x_0-y)}{|x_0-y|^n}\right| \leq Ck^{n/2}|x-x_0|/|x_0-y|^{n+1}$$

for $|x-y| > 2|x_0-x| > 0$. We will prove the following

$$(S_{hk}^{\tilde{A}}(f))^{\#}(\tilde{x}) \leq C\prod_{j=1}^{l}\left(\sum_{|\alpha_j|=m_j}||D^{\alpha_j}A_j||_{BMO}\right)k^{n/2}M_q(f)(\tilde{x}).$$

To do this, we write

$$S_{hk}^{\tilde{A}}(f_2)(x) - S_{hk}^{\tilde{A}}(f_2)(x_0) = \int_{R^n}\left(\frac{Y_{hk}(x-y)}{|x-y|^{n+m}} - \frac{Y_{hk}(x_0-y)}{|x_0-y|^{n+m}}\right) \tag{2.35}$$

$$\times \prod_{j=1}^{2} R_{m_j}(\tilde{A}_j;x,y)f_2(y)dy \tag{2.36}$$

$$+\int_{R^n}(R_{m_1}(\tilde{A}_1;x,y)-R_{m_1}(\tilde{A}_1;x_0,y))\frac{R_{m_2}(\tilde{A}_2;x,y)}{|x_0-y|^{m+n}}Y_{hk}(x_0-y)f_2(y)dy \tag{2.37}$$

$$+\int_{R^n}(R_{m_2}(\tilde{A}_2;x,y)-R_{m_2}(\tilde{A}_2;x_0,y))\frac{R_{m_1}(\tilde{A}_1;x_0,y)}{|x_0-y|^{m+n}}Y_{hk}(x_0-y)f_2(y)dy \tag{2.38}$$

$$-\sum_{|\alpha_1|=m_1}\frac{1}{\alpha_1!}\int_{R^n}[\frac{R_{m_2}(\tilde{A}_2;x,y)(x-y)^{\alpha_1}}{|x-y|^{m+n}}Y_{hk}(x-y) \tag{2.39}$$

$$-\frac{R_{m_2}(\tilde{A}_2;x_0,y)(x_0-y)^{\alpha_1}}{|x_0-y|^{m}}Y_{hk}(x_0-y)]D^{\alpha_1}\tilde{A}_1(y)f_2(y)dy \tag{2.40}$$

$$-\sum_{|\alpha_2|=m_2}\frac{1}{\alpha_2!}\int_{R^n}[\frac{R_{m_1}(\tilde{A}_1;x,y)(x-y)^{\alpha_2}}{|x-y|^{m+n}}Y_{hk}(x-y) \tag{2.41}$$

$$-\frac{R_{m_1}(\tilde{A}_1;x_0,y)(x_0-y)^{\alpha_2}}{|x_0-y|^{m+n}}Y_{hk}(x_0-y)]D^{\alpha_2}\tilde{A}_2(y)f_2(y)dy \tag{2.42}$$

$$+\sum_{|\alpha_1|=m_1,\ |\alpha_2|=m_2}\frac{1}{\alpha_1!\alpha_2!}\int_{R^n}[\frac{(x-y)^{\alpha_1+\alpha_2}}{|x-y|^{m+n}}Y_{hk}(x-y) \tag{2.43}$$

$$-\frac{(x_0-y)^{\alpha_1+\alpha_2}}{|x_0-y|^{m+n}}Y_{hk}(x_0-y)]D^{\alpha_1}\tilde{A}_1(y)D^{\alpha_2}\tilde{A}_2(y)f_2(y)dy \tag{2.44}$$

$$= I_5^{(1)}+I_5^{(2)}+I_5^{(3)}+I_5^{(4)}+I_5^{(5)}+I_5^{(6)}. \tag{2.45}$$

By Lemma 2.1 and the following inequality (see [13])

$$|b_{Q_1}-b_{Q_2}| \le C\log(|Q_2|/|Q_1|)||b||_{BMO} \text{ for } Q_1 \subset Q_2,$$

we know that, for $x \in Q$ and $y \in 2^{i+1}\tilde{Q}\setminus 2^i\tilde{Q}$,

$$|R_m(\tilde{A};x,y)| \le C|x-y|^m \sum_{|\alpha|=m}(||D^\alpha A||_{BMO}+|(D^\alpha A)_{\tilde{Q}(x,y)}-(D^\alpha A)_{\tilde{Q}}|) \tag{2.46}$$

$$\le Ci|x-y|^m \sum_{|\alpha|=m}||D^\alpha A||_{BMO}. \tag{2.47}$$

Note that $|x-y| \sim |x_0-y|$ for $x \in Q$ and $y \in R^n \setminus \tilde{Q}$, we obtain

$$|I_5^{(1)}| \leq Ck^{n/2}\int_{R^n}\frac{|x-x_0|}{|x_0-y|^{m+n+1}}\prod_{j=1}^{2}|R_{m_j}(\tilde{A}_j;x,y)||f_2(y)|dy \tag{2.48}$$

$$\leq C\prod_{j=1}^{2}\left(\sum_{|\alpha|=m_j}||D^\alpha A_j||_{BMO}\right)k^{n/2}\sum_{i=0}^{\infty}\int_{2^{i+1}\tilde{Q}\setminus 2^i\tilde{Q}}i^2\frac{|x-x_0|}{|x_0-y|^{n+1}}|f(y)|dy \tag{2.49}$$

$$\leq C\prod_{j=1}^{2}\left(\sum_{|\alpha|=m_j}||D^\alpha A_j||_{BMO}\right)k^{n/2}\sum_{i=1}^{\infty}i^2 2^{-i}\frac{1}{|2^i\tilde{Q}|}\int_{2^i\tilde{Q}}|f(y)|dy \tag{2.50}$$

$$\leq C\prod_{j=1}^{2}\left(\sum_{|\alpha|=m_j}||D^\alpha A_j||_{BMO}\right)k^{n/2}M(f)(\tilde{x}). \tag{2.51}$$

For $I_5^{(2)}$, by the formula (see [5]):

$$R_m(\tilde{A};x,y)-R_m(\tilde{A};x_0,y)=\sum_{|\beta|<m}\frac{1}{\beta!}R_{m-|\beta|}(D^\beta\tilde{A};x,x_0)(x-y)^\beta$$

and Lemma 2.1, we have

$$|R_m(\tilde{A};x,y)-R_m(\tilde{A};x_0,y)| \leq C\sum_{|\beta|<m}\sum_{|\alpha|=m}|x-x_0|^{m-|\beta|}|x-y|^{|\beta|}||D^\alpha A||_{BMO},$$

thus

$$|I_5^{(2)}| \leq C\prod_{j=1}^{2}\left(\sum_{|\alpha|=m_j}||D^\alpha A_j||_{BMO}\right)k^{n/2}\sum_{i=0}^{\infty}\int_{2^{i+1}\tilde{Q}\setminus 2^i\tilde{Q}}i\frac{|x-x_0|}{|x_0-y|^{n+1}}|f(y)|dy \tag{2.52}$$

$$\leq C\prod_{j=1}^{2}\left(\sum_{|\alpha|=m_j}||D^\alpha A_j||_{BMO}\right)k^{n/2}M(f)(\tilde{x}). \tag{2.53}$$

Similarly,

$$|I_5^{(3)}| \leq C\prod_{j=1}^{2}\left(\sum_{|\alpha|=m_j}||D^\alpha A_j||_{BMO}\right)k^{n/2}M(f)(\tilde{x}).$$

For $I_5^{(4)}$, we get

$$|I_5^{(4)}| \leq C\sum_{|\alpha_1|=m_1}\int_{R^n}\left|\frac{(x-y)^{\alpha_1}Y_{hk}(x,y)}{|x-y|^{m+n}}-\frac{(x_0-y)^{\alpha_1}Y_{hk}(x_0,y)}{|x_0-y|^{m+n}}\right| \tag{2.54}$$

$$\times|R_{m_2}(\tilde{A}_2;x,y)||D^{\alpha_1}\tilde{A}_1(y)||f_2(y)|dy \tag{2.55}$$

$$+C\sum_{|\alpha_1|=m_1}\int_{R^n}|R_{m_2}(\tilde{A}_2;x,y)-R_{m_2}(\tilde{A}_2;x_0,y)| \tag{2.56}$$

$$\times\frac{|(x_0-y)^{\alpha_1}Y_{hk}(x_0,y)|}{|x_0-y|^{m+n}}|D^{\alpha_1}\tilde{A}_1(y)||f_2(y)|dy \tag{2.57}$$

$$\leq C\sum_{|\alpha|=m_2}||D^{\alpha}A_2||_{BMO}k^{n/2} \tag{2.58}$$

$$\times\sum_{|\alpha_1|=m_1}\sum_{i=1}^{\infty}i2^{-i}\left(\frac{1}{|2^i\tilde{Q}|}\int_{2^i\tilde{Q}}|D^{\alpha_1}\tilde{A}_1(y)|^{q'}dy\right)^{1/q'}\left(\frac{1}{|2^i\tilde{Q}|}\int_{2^i\tilde{Q}}|f(y)|^{q}dy\right)^{1/q} \tag{2.59}$$

$$\leq C\prod_{j=1}^{2}\left(\sum_{|\alpha|=m_j}||D^{\alpha}A_j||_{BMO}\right)k^{n/2}M_q(f)(\tilde{x}). \tag{2.60}$$

Similarly,

$$|I_5^{(5)}|\leq C\prod_{j=1}^{2}\left(\sum_{|\alpha|=m_j}||D^{\alpha}A_j||_{BMO}\right)k^{n/2}M_q(f)(\tilde{x}).$$

For $I_5^{(6)}$, taking $r_1,r_2>1$ such that $1/q+1/r_1+1/r_2=1$, then

$$|I_5^{(6)}| \leq C\sum_{|\alpha_1|=m_1,|\alpha_2|=m_2}\int_{R^n}\left|\frac{(x-y)^{\alpha_1+\alpha_2}Y_{hk}(x,y)}{|x-y|^{m+n}}-\frac{(x_0-y)^{\alpha_1+\alpha_2}Y_{hk}(x_0,y)}{|x_0-y|^{m+n}}\right| \tag{2.61}$$

$$\times|D^{\alpha_1}\tilde{A}_1(y)||D^{\alpha_2}\tilde{A}_2(y)||f_2(y)|dy \tag{2.62}$$

$$\leq Ck^{n/2}\sum_{|\alpha_1|=m_1,|\alpha_2|=m_2}\sum_{i=1}^{\infty}i2^{-i}\left(\frac{1}{|2^i\tilde{Q}|}\int_{2^i\tilde{Q}}|f(y)|^{q}dy\right)^{1/q} \tag{2.63}$$

$$\times\left(\frac{1}{|2^i\tilde{Q}|}\int_{2^i\tilde{Q}}|D^{\alpha_1}\tilde{A}_1(y)|^{r_1}dy\right)^{1/r_1}\left(\frac{1}{|2^i\tilde{Q}|}\int_{2^i\tilde{Q}}|D^{\alpha_2}\tilde{A}_2(y)|^{r_2}dy\right)^{1/r_2} \tag{2.64}$$

$$\leq C\prod_{j=1}^{2}\left(\sum_{|\alpha|=m_j}||D^{\alpha}A_j||_{BMO}\right)k^{n/2}M_q(f)(\tilde{x}). \tag{2.65}$$

Thus

$$(S_{hk}^{\tilde{A}}(f))^{\#}(\tilde{x})\leq C\prod_{j=1}^{l}\left(\sum_{|\alpha_j|=m_j}||D^{\alpha_j}A_j||_{BMO}\right)k^{n/2}M_q(f)(\tilde{x})$$

and

$$||S_{hk}^{\tilde{A}}(f)||_{L^p(w)} \leq ||(S_{hk}^{\tilde{A}}(f))^{\#}||_{L^p(w)} \quad (2.66)$$

$$\leq C\prod_{j=1}^{l}\left(\sum_{|\alpha_j|=m_j}||D^{\alpha_j}A_j||_{BMO}\right)k^{n/2}||M_q(f)||_{L^p(w)} \quad (2.67)$$

$$\leq C\prod_{j=1}^{l}\left(\sum_{|\alpha_j|=m_j}||D^{\alpha_j}A_j||_{BMO}\right)k^{n/2}||f||_{L^p(w)}, \quad (2.68)$$

$$||I_5(\cdot)||_{L^p(w)} = \left|\left|\sum_{k=1}^{\infty}\sum_{h=1}^{g_k}a_{hk}S_{hk}^{\tilde{A}}(f)\right|\right|_{L^p(w)} \quad (2.69)$$

$$\leq C\prod_{j=1}^{2}\left(\sum_{|\alpha_j|=m_j}||D^{\alpha_j}A_j||_{BMO}\right)\sum_{k=1}^{\infty}\sum_{h=1}^{g_k}k^{-2n}||S_{hk}^{\tilde{A}}(f)||_{L^p(w)} \quad (2.70)$$

$$\leq C\prod_{j=1}^{2}\left(\sum_{|\alpha_j|=m_j}||D^{\alpha_j}A_j||_{BMO}\right)\sum_{k=1}^{\infty}k^{-2n+n/2+n-2}||f||_{L^p(w)} \quad (2.71)$$

$$\leq C\prod_{j=1}^{2}\left(\sum_{|\alpha_j|=m_j}||D^{\alpha_j}A_j||_{BMO}\right)||f||_{L^p(w)}. \quad (2.72)$$

These complete the proof of Theorem 1.3.

Proof of Theorem 1.4. Let $f \in L^{p,\varphi}(w)$. For a ball $B = B(x,d) \subset R^n$, note that $M(w\chi_B) \in \Lambda_1$(see Lemma 2.2), we get, by Theorem 1.3,

$$\int_B |T^A(f)(y)|^p w(y)dy = \int_{R^n}|T^A(f)(y)|^p w(y)\chi_B(y)dy \quad (2.73)$$

$$\leq \int_{R^n}|T^A(f)(y)|^p M(w\chi_B)(y)dy \quad (2.74)$$

$$\leq \int_{R^n}|f(y)|^p M(w\chi_B)(y)dy \quad (2.75)$$

$$= C\left[\int_B |f(y)|^p M(w\chi_B)(y)dy + \sum_{k=0}^{\infty}\int_{2^{k+1}B\setminus 2^kB}|f(y)|^p M(w\chi_B)(y)dy\right] \quad (2.76)$$

$$\leq C\left[\int_B |f(y)|^p w(y)dy + \sum_{k=0}^{\infty}\int_{2^{k+1}B\setminus 2^kB}|f(y)|^p \frac{w(B)}{|2^{k+1}B|}dy\right] \quad (2.77)$$

$$\leq C\left[\int_B |f(y)|^p w(y)dy + \sum_{k=0}^{\infty}\int_{2^{k+1}B}|f(y)|^p \frac{M(w)(y)}{2^{n(k+1)}}dy\right] \quad (2.78)$$

$$\leq C\left[\int_B |f(y)|^p w(y)dy + \sum_{k=0}^{\infty}\int_{2^{k+1}B}|f(y)|^p \frac{w(y)}{2^{nk}}dy\right] \quad (2.79)$$

$$\leq C||f||_{L^{p,\varphi}(w)}^p\sum_{k=0}^{\infty}2^{-nk}\varphi(2^{k+1}d) \quad (2.80)$$

$$\leq C||f||_{L^{p,\varphi}(w)}^p\sum_{k=0}^{\infty}(2^{-n}D)^k\varphi(d) \quad (2.81)$$

$$\leq C||f||_{L^{p,\varphi}(w)}^p\varphi(d), \quad (2.82)$$

thus,

$$||T^A(f)||_{L^{p,\varphi}(w)} \leq C||f||_{L^{p,\varphi}(w)}.$$

This completes the proof.

References

[1] A. P.Calderón and A.Zygmund, On singular integrals with variable kernels. *Appl. Anal.*, **7** (1978), 221-238.

[2] F.Chiarenza, M. Frasca and P. Longo, Interior $W^{2,p}$-estimates for nondivergence elliptic equations with discontinuous coefficients. *Ricerche Mat.*, **40** (1991), 149-168.

[3] J.Cohen, A sharp estimate for a multilinear singular integral on R^n, *Indiana Univ. Math. J.*, **30** (1981), 693-702.

[4] J.Cohen and J.Gosselin, On multilinear singular integral operators on R^n, *Studia Math.*, **72** (1982), 199-223.

[5] J.Cohen and J.Gosselin, A BMO estimate for multilinear singular integral operators, *Illinois J. Math.*, **30** (1986), 445-465.

[6] R.Coifman and Y.Meyer, Wavelets, Calderón-Zygmund and multilinear operators, *Cambridge University Press*, Cambridge Studies in Advanced Math. 48, Cambridge, 1997.

[7] Y.Ding and S.Z.Lu, Weighted boundedness for a class rough multilinear operators, *Acta Math. Sinica*, **17** (2001), 517-526.

[8] G.Di FaZio and M. A. Ragusa, Commutators and Morrey spaces, *Boll. Un. Mat. Ital.*, **(7)5-A** (1991), 323-332.

[9] G.Di Fazio and M. A. Ragusa, Interior estimates in Morrey spaces for strong solutions to nondivergence form equations with discontinuous coefficients, *J. Func. Anal.*, **112** (1993), 241-256.

[10] J.Garcia-Cuerva and J.L.Rubio de Francia, Weighted norm inequalities and related topics, *North-Holland Math. 16*, Amsterdam, 1985.

[11] S.Z.Lu, D.C.Yang and Z.S.Zhou, Oscillatory singular integral operators with Calderón-Zygmund kernels, *Southeast Asian Bull. of Math.*, **23** (1999), 457-470.

[12] J. Peetre, On the theory of $L^{p,\lambda}$-spaces, *J. Func. Anal.*, **4** (1969), 71-87.

[13] E.M.Stein, Harmonic Analysis: real variable methods, orthogonality and oscillatory integrals, *Princeton Univ. Press*, Princeton NJ, 1993.

In: African Diaspora Mathematics Research Progress
Editor: Toka Diagana, pp. 13–30
ISBN 978-1-60456-204-0
© 2008 Nova Science Publishers, Inc.

Chapter 2

L'INTEGRATION PAR RAPPORT A UNE MULTIMESURE, MONOTONE ET S-COMPACTE, A VALEURS CONVEXES FERMÉES

Gabriel Birame Ndiaye*
Départment de Mathématiques-Informatique, Université Cheikh Anta Diop,
Dakar, BP 5005, Sénégal

Abstract

We construct an integral relatively to a closed convex-valued multilinear measure. Our method unifies all the different integrals constructed by both D. S. Thiam and Pallu De La Barrière. For that, we introduce the notion of s-compact, which is for set-valued measures, what σ-finite is for scalar measures. We introduce also a notion of negligible functions. We apply these notions to the construction of the integral. We introduce a topology of the convergence in mean, for which, the spaces of integrable functions are complete.

AMS Subject Classification: 28B20; 28C05; 28B05.

Keywords: analyse convexe, intégration, multimesures, intégrale de Daniell.

1 Introduction

Les multimesures ont été le sujet de plusieurs thèses dont celle de C. Godet-Thobie [4] de l'école de C. Castaing, et celles de l'école de Pallu De La Barrière : D. S. Thiam, A. Costé, K. Siggini. Le point de vue des semi-variations a été apporté par Pallu De La Barrière [9]. Tous ces travaux ont été fait pour une multimesure à valeurs faiblement compactes. D. S. Thiam a étendue la théorie de l'intégrale de Daniell aux multimesures. Entre autres, deux approches y sont présentées : celle de l'intégrale de Daniell secondaire, et celle de l'intégrale multivoque. Il a aussi effectué une tentative d'intégration par rapport à une multimesure à valeurs convexes fermées bornées. Ce sont donc trois méthodes d'intégration qu'il a présenté, sans aucun lien entre elles. Nous avons eu le projet d'étudier l'intégration

*E-mail addresses: gabrielbirame@ucad.sn, gabrielbirame@yahoo.fr

par rapport à une multimesure à valeurs convexes fermées. Cela nous a amené à définir dans [5], la notion de s-compacité, qui est l'équivalent multivoque de σ-finie des mesures scalaires, et avions obtenu différents résultats, parmi lesquels, les théorèmes de convergence au sens de l'ordre, et une caractérisation simple et pratique de l'intégrabilité d'une fonction. Un des objectifs de cet article est, d'unifier les trois méthodes de [12], et de faire le lien avec [9], tout en apportant des simplifications dans les démonstrations. Cet article est structuré comme suit. Dans la section 2, nous rappelons des définitions et des résultats de [5] et [7], et nous construisons sur le modèle de [12], une topologie dans l'espace des fonctions dont l'intégrale supérieure est bornée, espace qui contient celui de [5]. Nous définissons ensuite un espace de fonctions intégrables, adhérence de l'ensemble des fonctions étagées, comme dans [12], mais cette fois ci dans le cas non compact. Dans la section 3, nous définissons la notion de fonctions et d'ensembles négligeables, avec une approche différente de [9], ce qui permet de généraliser des résultats de [5]. Dans la section 4, nous définissons la topologie de la convergence en moyenne. Nous établissons que les espaces de fonctions intégrables sont complets, contiennent comme sous-espaces topologiques ceux de [12], [9], et que de toute suite convergente, on peut extraire une sous-suite qui converge presque-partout. Ce théorème, hormis l'unification de différents travaux, est le résultat principal de cet article. Nous obtenons également les théorèmes de convergences en moyenne, qui généralisent ceux de [12]. Dans la section 5, nous prouvons que nos espaces de fonctions intégrables, contiennent strictement ceux obtenus par [12]. Nous vérifions aussi que l'ensemble des convexes faiblement compacts, d'un espace non complet, n'est pas toujours fermé dans celui des convexes fermés bornés, et donc intégrer comme dans [9], nécessite que l'espace soit complet. Dans la section 6, nous généralisons des résultats de [12], pour une mesure vectorielle, grâce aux nouveaux outils que nous venons de définir.

2 Notations et Préliminaires

On considére E un espace topologique localement convexe séparé non complet, E' son dual topologique. Le crochet de dualité $\langle E, E' \rangle$ est $\langle \cdot, \cdot \rangle$. Pour $A \subset E$, et $y \in E'$, on pose $\delta^*(y/A) = \sup\{\langle x, y \rangle : x \in A\}$. On considère les monoïdes suivants : $(cc(E), +)$ l'ensemble des convexes faiblement compacts de E, $(cfb(E), \dot{+})$ celui des convexes fermés bornés non vides, et $(cf(E), \dot{+})$ celui des convexes fermés non vides. Si $A, B \in cf(E)$, $\overline{A}$ est l'adhérence de A et, $A \dot{+} B = \overline{(A+B)}$. On désigne par $\overline{co}(A)$, l'enveloppe convexe fermée de l'ensemble A. La structure uniforme, de ces monoïdes, est celle de Hausdorff [1], et la relation d'ordre l'inclusion. Si O est une famille de parties convexes fermées de E on a : $\inf\{A : A \in O\} = \cap\{A : A \in O\}$, et $\sup\{A : A \in O\} = \overline{co} \cup \{A : A \in O\}$. Soient $(A_n)_{n \in \mathbb{N}}$, une suite de cf(E). On note : $\overline{\lim} A_n = \inf\limits_{n \in \mathbb{N}} \sup\limits_{p \geq n} A_n$, et $\underline{\lim} A_n = \sup\limits_{n \in \mathbb{N}} \inf\limits_{p \geq n} A_n$, au sens de l'ordre dans cf(E). Si en plus, pour $A \in cf(E)$, on a $A = \overline{\lim} A_n = \underline{\lim} A_n$, alors A est la limite, au sens de l'ordre, dans cf(E), de la suite $(A_n)_{n \in \mathbb{N}}$. On définit de manière analogue, la limite au sens de l'ordre dans cfb(E), et dans cc(E). On désigne par T un ensemble non vide. Si $f \in \overline{\mathbb{R}}^T$ est une fonction de T dans $\overline{\mathbb{R}}$, on pose $f^+ = \sup(f, 0)$, et $f^- = \sup(-f, 0)$. Si $A, B \in cfb(E)$, et si E est normé, on désigne par DH(A, B) la distance de Hausdorff. La formule de Hörmander(voir [1], chapitre II) sécrit :

$$DH(A,B) = \sup\{|\delta^*(y/A)) - \delta^*(y/B))| : y \in E', \|y\| \leq 1\}.$$

Soit M: $\Omega \longrightarrow$ cf(E) une multiapplication. On dira que M est une multimesure faible si $\delta^*(y/M(.))$ est une mesure scalaire pour tout y de E' avec M(∅)= $\{0\}$: cette notion a étée introduite dans [2].

Définition 2.1. On considère une multimesure faible M à valeurs dans cf(E).

(1) On dit que M est séquentiellement compacte (s-compacte), s'il existe une suite $(T_n)_{n\in\mathbb{N}^*}$ de terme général $T_n \in \Omega$, croissante vérifiant $\cup_{n\in\mathbb{N}^*} T_n = T$ (on note $T_n \uparrow T$) et telle que, pour tout $n \in \mathbb{N}^*, M(T_n) \in cc(E)$.

(2) On dit que M est monotone si on a : $0 \in M(A)$, $\forall A \in \Omega$.

Exemple 2.2. Soit $\mathbb{B}(\mathbb{R}^n)$ la tribu borélienne de $\mathbb{R}^n$ et λ une mesure σ-finie sur $\mathbb{B}(\mathbb{R}^n)$. On considére : $M : \mathbb{B}(\mathbb{R}^n) \longrightarrow cf(\mathbb{R})$, définie par : $M(A) = [0, \lambda(A)]$, si $\lambda(A) < +\infty$, et $M(A) = [0, +\infty[$, sinon. M est une multimesure monotone s-compacte. L'additivité provient de la propriétée de décomposition de Riesz, cf par exemple [12] page 173.

On considère une multimesure faible M à valeurs dans cf(E), et on suppose que M est s-compacte et monotone. On pose : $\Lambda = \{A \in \Omega / M(A) \in cc(E)\}$. Si h= $\Sigma a_i 1_{A_i}$ est une fonction en escaliers basées sur le clan Λ, soit $I(h) = \Sigma a_i M(A_i)$, avec h positive. Si on désigne par H l'espace de Riesz des fonctions en escaliers basées sur Λ, et $H_+ = \{h \in H : h \geq 0\}$, alors l'application $I : H_+ \to cc(E)$ est une intégrale de Daniell secondaire : I est additive, monotone i.e. $0 \in I(h)$ pour tout $h \in H_+$, et si $(h_n)_{n\in\mathbb{N}^*}$ est une suite décoissante de H_+, qui converge simplement vers 0, (on note $h_n \downarrow 0$), alors $\inf_{n\in\mathbb{N}^*} I(h_n) = \{0\}$. Ici on désigne par $H^\vee$ (resp $H^\wedge$) l'ensemble des fonctions qui sont limites croissante (resp décroissante) de fonctions de H_+. Pour $h_n \uparrow f$, $h_n \in H_+$ et $g_n \downarrow g$, $g_n \in H_+$, on pose : $I^\vee(f) = \sup_{n\in\mathbb{N}^*} I(h_n)$, et $I^\wedge(g) = \inf_{n\in\mathbb{N}^*} I(g_n)$. Si $f \in \overline{\mathbb{R}}_+^T$, on pose : $I^*(f) = \inf\{I^\vee(\psi), \psi \in H^\vee : \psi \geq f\}$, si cet ensemble est non vide, sinon on pose $I^\star(f) = E$, et $I_*(f) = \sup\{I^\wedge(\phi), \phi \in H^\wedge, \phi \leq f\}$. L'espace des fonctions intégrables pour l'intégrale de Daniell scondaire I est :

$$L_+^1(I) = \{f \in \overline{\mathbb{R}}_+^T, \ I^*(f) = I_*(f) \in cfb(E)\}.$$

C'est un cône convexe réticulé, voir [5] ou [7]. Si $f \in L_+^1(I)$, $\int fI = I^*(f)$ et on le notera $\int(.)I$ ou I. Elle est additive et positivement homogène. On posera :

$$\text{si } h \in H_+, I_y(h) = \delta^*(y, I(h)), \text{ et si } h \in H, I_y(h) = I_y(h^+) - I_y(h^-).$$

Nous rappellons la propriétée suivante : si $(C_n)_{n\in\mathbb{N}^*}$ est une suite décroissante de cc(E), alors pour tout $y \in E'$ on a :

$$\inf_{n\in\mathbb{N}^*} \delta^*(y/C_n) = \delta^*(y/ \inf_{n\in\mathbb{N}^*} C_n) \tag{2.1}$$

Dans la remarque 2.3 suivante, nous allons essayer de situer la position du problème.

Remarque 2.3. L'approche que nous avons expoxé, ci avant, est celle de [12]. Cependant pour assurer par exemple la sous-additivitée de I^* : $I^*(f_1+f_2) \subset I^*(f_1)\dot{+}I^*(f_2)$, il fallait supposer que $I^*(f_1) \in cc(E)$, $I^*(f_2) \in cc(E)$. En effet, cette sous-additivitée est basée sur la propriétée 2.1. Or dans cfb(E), cette propriétée ne subsiste plus : à la page 78 de [12], on peut trouver un contre- exemple, communiqué par le professeur Michel Valadier. Donc dans cfb(E), la construction s'effondre. C'est pourquoi, pour une multimesure faible à valeurs dans cfb(E), [12] a utilisé une autre approche, comme spécifié dans notre remarque 4.10. Si maintenant la multimesure est à valeurs dans cf(E), comme dans l'exemple 2.2, que peut on faire ? L'exemple 2.2 nous a suggéré la notion de séquentielle compacité (s-compact). Elle nous a permis de maintenir l'édifice, et d'obtenir, en particulier, dans [5], les différents résultats résumés dans le théorème 2.5, et la proposition 2.4. Ces mêmes résultats avaient été obtenus dans [12], mais en supposant que tout ce passe dans cc(E). Ici notre multimesure est à valeurs dans cf(E). De plus l' intégrale d'une fonction peut être à valeurs dans cfb(E), et non plus seulement dans cc(E), comme dans [9] ou [12]. D'autre part, la notion de négligeable, que nous introduisons à la section 3, n'existait pas dans [5], ni dans [12]. Tous les résultats obtenus Dans [12], le sont essentiellement, pour des fonctions définies sur T à valeurs dans $\mathbb{R}$. La même situation se pose dans l'approche présenté, après le théorème 2.5. Au total, on a eu dans [12], trois méthodes différentes d'intégration. Le lien entre ces méthodes n'avait pas encore été établi, à notre connaissance.

La proposition 2.4 suivante regoupe dans l'ordre les propositions 7 et 13, puis les théorèmes 6 et 1 que nous avions obtenu dans [5].

Proposition 2.4. *On considère* $f \in \overline{\mathbb{R}}^T$, *et* $y \in E'$.

(1) On a $(I_y)^*(|f|)=\delta^*(y/I^*(|f|))$.

(2) Si $I_*(|f|) \in cfb(E)$, *alors on obtient,* $(I_y)_*(|f|) = \delta^*(y/I_*(|f|))$.

(3) On a aussi $L^1_+(I) = \cap_{y\in E'} L^1_+(I_y)$, *et si* $f \in L^1_+(I)$ *alors,* $\delta^*(y/\int fI) = \int f\delta^*(y/I(.))$.

(4) Soit $(f_n)_{n\in\mathbb{N}^*}$ *une suite croissante de* $\overline{\mathbb{R}}_+^T$ *on a,* $\sup_{n\in\mathbb{N}^*} I^*(f_n) = I^*\left(\sup_{n\in\mathbb{N}^*}(f_n)\right)$.

Dans [5] nous avions obtenu, comme dans [12], les théorèmes de convergences monotones et dominée, au sens de l'ordre, mais cette fois ci, pour une multimesure à valeurs convexes fermées de E, et pour des fonctions à valeurs dans $\overline{\mathbb{R}}$:

Théorème 2.5. *(1) Si* $(f_n)_{n\in\mathbb{N}^*}$ *est une suite croissante de* $L^1_+(I)$, *telle que* $\sup_{n\in\mathbb{N}^*}\int f_nI \in$ *cfb(E), alors :* $\sup_{n\in\mathbb{N}^*} f_n \in L^1_+(I)$, *et* $\int \sup_{n\in\mathbb{N}^*} f_nI = \sup_{n\in\mathbb{N}^*}\int f_nI$.

(2) Soit $(f_n)_{n\in\mathbb{N}^*}$ *une suite décroissante de* $L^1_+(I)$, *on a :* $\inf_{n\in\mathbb{N}^*} f_n \in L^1_+(I)$ *et* $\int \inf_{n\in\mathbb{N}^*} f_nI = \inf_{n\in\mathbb{N}^*}\int f_nI$.

(3) Soit $(f_n)_{\in\mathbb{N}^*}$ *une suite de* $L^1_+(I)$, *telle qu'il existe* $g \in \overline{\mathbb{R}}_+^T$, *avec* $I^*(g) \in cfb(E)$, *et* $f_n \leq g$ *pour* $n \geq 1$. *Si* $(f_n)_{n\in\mathbb{N}^*}$ *converge simplement vers f on a :* $f \in L^1_+(I)$, *et* $\int fI = \lim_{n\to+\infty}\int f_nI$ *au sens de l'ordre.*

Nous appliquons une méthode de [12] pour l'intégrale multivoque, mais cette fois ci dans le cas non compact. Si on pose : $H^* = \{f \in \mathbb{R}^T,\ I^*(|f|) \in cfb(E)\}$, alors $(H^*,+,\leq)$ est un sous espace de Riesz de $\mathbb{R}^T$. Soit $V(0)$ l'ensemble des voisinages de 0 dans E ; si $V \in V(0)$, posons : $V^* = \{f \in \mathbb{R}^T, I^*(|f|) \subset V\}$. L'ensemble $V^*(0)$ des V^* quand $V \in V(0)$ est une base de filtre de 0 de H^*. Nous avons ainsi défini sur H^* une topologie faisant de H^* un espace localement convexe. l'espace $L^1_+(I)$, que nous avons défini dans [5] ou [7], est l'espace des fonctions intégrables pour l'intégrale de Daniell secondaire, où I est l'intégrale associée à la multimesure M. Ainsi, on a : $H_+ \subset L^1_+(I) \cap \mathbb{R}^T \subset H^*$. Nous venons donc de munir d'une topologie l'espace $L^1_+(I) \cap \mathbb{R}^T$ et ceci n'existait pas dans [12]. Nous en ferons de même pour $L^1_+(I)$. Ces espaces contiennent ceux obtenus par [12], et nous verrons, à la section 5, que l'inclusion peut même être stricte. Dans [12], la sous-additiviée de I^* n'étant pas assuré dans cfb(E), cette construction n'était possible que dans cc(E).

Définition 2.6. On désigne par $\ell^1(I)$ l'adhérence de H dans H^*.

Proposition 2.7. *On a* $\ell^1(I) = \ell^1_+(I)$ - $\ell^1_+(I)$, *et* $\ell^1_+(I) = \overline{H}_+$ *dans* H^*_+.

Preuve. a) Soient $f_1, f_2 \in \ell^1(I)$ et $V \in V(0)$. Considérons un élément W de V(0) tel que $W + W \subset V$. Il existe $h_1, h_2 \in H$: $I^*(|f_i - h_i| \subset W$, pour i=1, 2. Remarquons que :

$$I^*(|\sup(f_1,f_2) - \sup(h_1,h_2)|) \subset I^*(|f_1 - h_1|) + I^*(|f_2 - h_2|),\ \text{et}$$

$$I^*(|\inf(f_1,f_2) - \inf(h_1,h_2)|) \subset I^*(|f_1 - h_1|) + I^*(|f_2 - h_2|),\ \text{donc,}$$

$$I^*(|\sup(f_1,f_2) - \sup(h_1,h_2)|) \subset I^*(|f_1 - h_1|) + I^*(|f_2 - h_2|) \subset W + W \subset V,\ \text{et}$$

$$I^*(|\inf(f_1,f_2) - \inf(h_1,h_2)|) \subset I^*(|f_1 - h_1|) + I^*(|f_2 - h_2|) \subset V.$$

Puisque H est réticulé, $\sup(h_1,h_2)$ et $\inf(h_1,h_2)$ sont dans H. De même H^* est réticulé, donc $\sup(f_1,f_2)$ et $\inf(f_1,f_2)$ sont dans H^*. Ainsi $\ell^1(I)$ est réticulé et par suite $\ell^1(I) = \ell^1_+(I) - \ell^1_+(I)$. b) Il est clair que $\overline{H}_+ \subset \ell^1_+(I)$. Soit $f \in \ell^1_+(I)$, $\exists h \in H$: $I^*(|f - h| \subset V$. Comme $|f - h_+| \leq |f - h|$, on a $I^*(|f - h_+|) \subset I^*(|f - h|) \subset V$. Ainsi $f \in \overline{H}_+$, et donc on a $\ell^1_+(I) \subset \overline{H}_+$. □

Proposition 2.8. *Nous avons les assertions suivantes :*

(1) $[f_n \to f$ *dans* $H^*] \Leftrightarrow [I^*(|f_n - f| \to 0$ *dans cfb(E)]* $\Rightarrow [I^*(|f_n| \to I^*|f|$ *dans cfb(E],*

(2) $[f_n \to f$ *dans* $L^1_+(I) \cap \mathbb{R}^T] \Leftrightarrow [f_n, f \in L^1_+(I) \cap \mathbb{R}^T$ *et* $I^*(|f_n - f| \to 0$ *dans* $cfb(E]$,

(3) Si $L^1_+(I) \cap \mathbb{R}^T$ *est muni de la topologie induite par celle de* H^*_+ *et cfb(E) de la topologie de Hausdorff, alors l'application I, I :* $L^1_+(I) \cap \mathbb{R}^T \to$ *cfb(E), est uniformément continue.*

Preuve. Preuve de (1) : l'assertion suivante : $\forall\ V \in V(0), \exists N, \forall n \geq N, I^*(|f_n - f|) \in V$, signifie que $I^*(|f_n - f|$ tend vers 0 dans cfb(E). On en déduit la première équivalence. Pour l'implication, nous avons : $|f_n| \leq |f| + |f_n - f|$ *et* $|f| \leq |f_n| + |f_n - f|$. Donc si $I^*(|f_n - f|) \in V$, on a : $I^*(|f_n|) \leq I^*(|f|) + V$, *et* $I^*(|f|) \leq I^*(|f_n|) + V$. Le (2) résulte du (1), car la topologie de $L^1_+(I) \cap \mathbb{R}^T$ est induite par celle de H^*. (3) Soient $f, f' \in L^1_+(I) \cap \mathbb{R}^T$ tels que $I^*(|f - f'| \subset V$, vu que $I(f) \subset I(f') + I(|f - f'|)$ et $I(f') \subset I(f) + I(|f - f'|$, on a $I(f) \subset I(f') + V$ et $I(f') \subset I(f) + V$. □

Remarque 2.9. Si f $\in \ell^1_+(I)$, il existe une suite généralisée de H_+ qui converge vers f. On posera I(f)=lim (I(h_α) $= I^*(f)$. Si f$\in \ell^1(I)$, on pose I(f)=I(f^+)$\dot{+}(-I(f^-))$, où : (- I(f^-)= { $(-x)$$\in$E,x$\in$ $I(f^-)$ }. Si E est complet, d'après la proposition 2 page I-15 de [8], cc(E) est fermé dans cf(E). Donc si f$\in \ell^1(I)$, alors I(f)$\in$cc(E), ie que l'espace $\ell^1(I)$ devient l'espace obtenu par [12] pour l'intégrale multivoque : cela n'est plus le cas si E non complet, voir section 5.

3 Fonctions et Ensembles I-négligeables

Définition 3.1. Une fonction f de $\overline{\mathbb{R}}^T$ est dite I-négligeable si $I^*(|f|) = \{0\}$. On dit qu'un sous-ensemble A de T est I-négligeable si 1_A est I-négligeable.

Proposition 3.2. *(1) Une partie d'un ensemble I-négigeable est I-négligeable.*

(2) Une réunion dénombrable d'ensembles I-négligeables est I-négigeable.

Preuve. (1) Si on a B$\subset$A, alors $I^*(1_B)\subset I^*(1_A)$ ={ 0 }. (2) Soit $(A_n)_{n\in\mathbb{N}^*}$ une suite de parties négligeables de T. Posons : $B_n = \cup_{i=1}^{i=n} A_i$, *on a* $B_n \uparrow \cup_{i=1}^{i=+\infty} A_i$ *et* $\sup_n 1_{B_n} = 1_{\cup_{i=1}^{i=+\infty} A_i}$.

D'après la proposition 2.4 (4) on a : $\sup_n I^*(1_{B_n}) = I^*(1_{\cup_{i=1}^{i=+\infty} A_i})$; or $0 \le 1_{B_n} \le \sum_{i=1}^{i=n} A_i$, et comme I^* est sous - additive on a le résultat. □

Définition 3.3. On dit qu'une propriétée p(t) est vraie I-presque-partout (I-pp ou pp) par rapport à I, si l'ensemble où [non p(t)] est vraie est I-négligeable.

Proposition 3.4. *On a les assertions suivantes :*

(i) f est I-négligeable si et seulement si $|f| \in L^1_+(I)$ *et* $\int |f| I = \{\ 0\}$,

(ii) $f \in \overline{\mathbb{R}}^T$ *est I-négligeable si et seulement si f = 0 I-pp,*

(iii) Si $I^*(|f|) \in$ *cfb(E) alors f est finie I-pp.*

Preuve. Preuve de (i) et (ii) : f est I-négligeable $\Leftrightarrow I^*(|f|) = \{\ 0\} \Leftrightarrow |f| \in L^1_+(I)$ et $\int |f| I = \{\ 0\}$ d'où (i). De plus d'après la proposition 2.4 (3), cela est équivalent à : $|f| \in L^1_+(I_y)$ et $\int |f|\ I_y = 0 \forall y \in E'$ $\Leftrightarrow |f|$ est I_y-négligeable pour tout y de E' $\Leftrightarrow f = 0\ I_y - pp\ \forall y \in E' \Leftrightarrow f = 0\ I - pp$, i.e. (ii). Preuve de (iii) : D'après la proposition 2.4 (1), $I^*(|f|) \in cfb(E) \Leftrightarrow (I_y)^*(|f|)$ est fini pour tout y de E'. Soit $N = \{\ t \in T, |f(t)| = +\infty\}$, On a $(I_y)^*(1_N) = 0\ \forall y \in E'$, ie $\delta^*(y/I^*(1_N) = 0\ \forall y \in E'$. On obtient donc que $I^*(1_N) = \{0\}$, i.e. que f est fini presque-partout. □

D'après cette preuve et du fait que $(I_y)^*(|f|)$=$\delta^*(y/I^*(|f|))$ on a la remarque suivante :

Remarque 3.5. On a, f est I-négligeable si et seulement si f est I_y-négligeable pour tout y de E'. Une propriétée p(t) est vraie I-pp si et seulement si p(t) est vraie I_y-pp pour tout y de E'.

Remarque 3.6. Si f$\in L^1_+(I)$ alors f est finie I-pp . Dans $\overline{\mathbb{R}}^T$ la relation f R g si et seulement si f = g I-pp est une relation d'équivalence.

Définition 3.7. Deux fonctions de $\overline{\mathbb{R}}^T$ sont équivalentes si elles sont égales presque partout. La relation R sera notée I-pp.

Proposition 3.8. *Soient f, g* $\in \overline{\mathbb{R}}^T$ *avec f et g positives on a : Si f=g I-pp alors* $I^*(f) = I^*(g)$. *De plus Si f=g I-pp et si* $I_*(f)$, $I_*(g) \in$ *cfb(E) alors* $I_*(f) = I_*(g)$.

Preuve. On se ramène au cas scalaire avec la proposition 2.4 (1) et (2). □

Définition 3.9. Soit N ⊂T où N est un I-négligeable. Une fonction définie sur T-N est une fonction définie I-pp sur T. Posons F(T, $\overline{\mathbb{R}}$) = { f définie I-pp sur T , $I^*(|f|)$ ∈cfb(E) } et Δ l'espace quotient. Δ est un espace vectoriel.

Définition 3.10. On pose $£^1_+(I) = \{f \in F(T,\overline{\mathbb{R}}),\ f \geq 0 \,/ \exists g \in L^1_+(I),\ f = g$ I-pp $\}$.

Le théorème 3.11 est une version I-pp des théorèmes de convergence de [5] : cf théorème 2.5. La notion de négligeable, que nous avons introduit, permet d'énoncer ces théorèmes, pour des fonctions définies I-pp sur T à valeurs dans $\overline{\mathbb{R}}$.

Théorème 3.11. *(1) Si* $(f_n)_{n\in\mathbb{N}^*}$ *est une suite croissante I-pp de* $£^1_+(I)$, *s'il existe* $g \in F(T,\overline{\mathbb{R}})$ *telle que* $f_n \leq g$ *I-pp pour tout* $n \in \mathbb{N}^*$, *et si on a* $I^*(g) \in cfb(E)$, *alors* $\sup_{n\in\mathbb{N}^*} f_n \in £^1_+(I)$ *et* $\int (\sup_{n\in\mathbb{N}^*} f_n) I = \sup_{n\in\mathbb{N}^*} \int f_n I.$

(2) Si $(f_n)_{n\in\mathbb{N}^*}$ *est une suite décroissante I-pp de* $£^1_+(I)$, *alors* $\inf_{n\in\mathbb{N}^*} f_n \in £^1_+(I)$ *et* $\int (\inf_{n\in\mathbb{N}^*} f_n) I = \inf_{n\in\mathbb{N}^*} \int f_n I.$

(3) Si $(f_n)_{n\in\mathbb{N}^*}$ *est une suite de* $£^1_+(I)$, *s'il existe* $g \in F(T,\overline{\mathbb{R}})$ *telle que* $f_n \leq g$ *I-pp pour tout* $n \in \mathbb{N}^*$, *avec* $I^*(g) \in cfb(E)$, *et si* $(f_n)_{n\in\mathbb{N}^*}$ *converge simplement vers f I-pp, alors* $f \in £^1_+(I)$ *et* $\int fI = \lim_{n\to+\infty} \int f_n I$ *au sens de l'ordre dans cfb(E).*

Preuve. on pose N_n={ t∈T , f_n non définie en t } N'_n={ t∈T, f_n non définie en t } ∪{ t /f_n(t) >g(t) } , $f(t) = \sup_n f_n(t)$, et N''_n={ t∈T, f_n ne croît pas vers f(t) } . Si t∈ $N_n \cup N'_n \cup N''_n$ qui est négligeable, on pose g_n(t)=0 et g_n(t) =f_n(t) sinon. On a $g_n \in L^1_+(I)$, et en prenant g=0 sur $N_n \cup N'_n \cup N''_n$, on a $g_n \leq$ g. On a $g_n \uparrow \sup_n g_n$ et $I^*(g_n) \subset I^*(g)$. On applique le théorème 2.5 de la convergence monotone : voir aussi [5] ou [7]. Comme $\sup_{n\in\mathbb{N}^*} f_n = \sup_{n\in\mathbb{N}^*} g_n$ I-pp, on a le résultat. La démonstration est similaire pour 2) et 3). □

4 Topologie de la Convergence en Moyenne

On suppose dans toute la suite que E est normé sauf spécification contraire. Pour f ∈F(T,$\overline{\mathbb{R}}$) si on pose :

$$N_1(f) = sup\{\delta^*(y/I^*(|f|)),\ y \in E', \|y\| \leq 1\} = DH(\{0\}, I^*(|f|)),$$

alors N_1 est une semi-norme sur F(T,$\overline{\mathbb{R}}$). De plus $N_1(f) = 0 \Leftrightarrow f = 0\ I-pp$. On en déduit, d'après la proposition 3.4 (ii), la remarque 4.1.

Remarque 4.1. On a : I-négligeable est équivalent à N_1-négligeable de [9], si nous prenons comme semi-variation N_1 .

Définition 4.2. On dit qu'une suite $(f_n)_{n\in\mathbb{N}^*}$ de F(T,$\overline{\mathbb{R}}$) converge en moyenne vers f si $\lim_{n\to+\infty} N_1(f_n - f) = 0$. La topologie de F(T,$\overline{\mathbb{R}}$), muni de la semi-norme N_1, est la topologie de la convergence en moyenne. L'application $\hat{f} \mapsto N_1(f)$, $f \in \hat{f}$, est une norme sur Δ.

Sauf spécification contraire, les espaces $F(T,\overline{\mathbb{R}})$ et Δ seront considérés comme munis de N_1.

Définition 4.3. On désigne par $\mathsf{L}^1(I)$ l'adhérence de H dans F(T,$\overline{\mathbb{R}}$) et on posera : $\mathsf{L}^1_+(I) = \{f \in \mathsf{L}^1(I),\ f \geq 0\}$.

NB: Notons la différence entre $\mathsf{L}^1_+(I)$ *et* $L^1_+(I)$.

Proposition 4.4. *(i) L'espace* $\mathsf{L}^1(I)$ *est réticulé et* $\mathsf{L}^1(I) = \mathsf{L}^1_+(I)$ - $\mathsf{L}^1_+(I)$.

(ii) On a aussi $\mathsf{L}^1_+(I) = \overline{H}_+$ *dans F(T,$\overline{\mathbb{R}}$).*

(iii) Si $\pounds^1_+(I)$ *est muni de la topologie de la convergence en moyenne et cfb(E) de celle de Hausdorff, alors l'application I :* $\pounds^1_+(I) \to$ *cfb(E) est uniformément continue.*

(iv) On a $\ell^1(I) \subset \mathsf{L}^1(I)$ *et,* $f \in \mathsf{L}^1(I) \Leftrightarrow \exists g \in \ell^1(I)$, $g = f$ *I-pp. La topologie de* $\ell^1(I)$ *est celle induite par* $\mathsf{L}^1(I)$.

Preuve. Remarquons que $0 \in I^*(|f|)$ pour tout f définie I-pp sur T à valeurs dans $\overline{\mathbb{R}}$, et $0 \leq f \leq g$ I-pp $\Rightarrow I^*(f) \subset I^*(g)$. (i) Soient $f_1, f_2 \in \mathsf{L}^1(I)$ et $\varepsilon \in \mathbb{R}^*_+$, il existe $h_1, h_2 \in H$ tels que $N_1(f_i - h_i) < \frac{\varepsilon}{2}$ pour i=1,2. Si $a,b,c,d \in \mathbb{R}^T$ on a $|\sup(a,b) - \sup(c,d)| \leq |a-c| + |b-d|$ et, $|\inf(a,b) - \inf(c,d)| \leq |a-c| + |b-d|$. Si a, b, c, d$\in F(T,\overline{\mathbb{R}})$, ces inégalitées ont lieu I-pp. Par conséquent : $I^*(|\sup(f_1,f_2) - \sup(h_1,h_2)|) \subset I^*(|f_1 - h_1|) + I^*(|f_2 - h_2|)$, donc

$$\forall y \in E', \delta^*(y/I^*(|\sup(f_1,f_2) - \sup(h_1,h_2)|)) \leq \delta^*(y/I^*(|f_1 - h_1|)) + \delta^*(y/I^*(|f_2 - h_2|)),$$

d'où : $N_1(\sup(f_1,f_2) - \sup(h_1,h_2)) \leq N_1(f_1 - h_1) + N_1(f_2 - h_2) < \varepsilon$. De Façon similaire, on obtient $N_1(\inf(f_1,f_2) - \inf(h_1,h_2)) \leq \varepsilon$. Ainsi on a $\sup(f_1,f_2), \inf(f_1,f_2) \in \mathsf{L}^1(I)$, donc $\mathsf{L}^1(I)$ est réticulé et par suite $\mathsf{L}^1(I) = \mathsf{L}^1_+(I)$ - $\mathsf{L}^1_+(I)$. (ii) Soient $f \in \mathsf{L}^1_+(I)$ et $\varepsilon \in R^*_+$, $\exists h \in H$: $N_1(f-h) < \varepsilon$. Or $|f - h^+| \leq |f-h|$ I-pp, et I^* est croissante, on en déduit $I^*(|f - h^+|) \subset I^*(|f-h|)$, par conséquent : $\forall y \in E'$, $\delta^*(y/I^*(|f - h^+|)) \leq \delta^*(y/I^*(|f-h|))$. On a alors $N_1(f - h^+) \leq \varepsilon$. Ainsi f appartient à l'adhérence de H_+ dans $F(T,\overline{\mathbb{R}})$. L'inclusion réciproque découlant immédiatement de la définition 4.3, nous avons l'égalité annoncée. (iii) Soient $f, f' \in \pounds^1_+(I)$, vu que $I(f) \subset I(f') + I(|f - f'|)$ et $I(f') \subset I(f) + I(|f - f'|$ alors on a : $\forall y \in E'$, $\delta^*(y/I(f)) \leq \delta^*(y/I(f')) + \delta^*(y/I(|f - f'|))$ et $\delta^*(y/I(f')) \leq \delta^*(y/I(f)) + \delta^*(y/I(|f - f'|)$. Ainsi on a : $\forall y \in E'$, $|\delta^*(y/I(f)) - \delta^*(y/I(f'))| \leq \delta^*(y/I(|f - f'|))$. On en déduit : $DH(I(f), I(f')) \leq N_1(f - f')$, d'où le résultat. (iv) Soit $V_\varepsilon = \{v \in E / \forall y \in E', \|y\| \leq 1 \Rightarrow \langle v,y \rangle < \varepsilon\} \equiv B(0,\varepsilon)$. On a : $V^*_\varepsilon = \{f \in \mathbb{R}^T\ , N_1(f) < \varepsilon\} \subset \{f \in F(T,\overline{\mathbb{R}}, N_1(f) < \varepsilon\}$. Cela signifie que, si E est normé, la topologie de H^* est celle induite par la topologie de $F(T,\overline{\mathbb{R}})$. L'assertion résulte alors des définitions de $\ell^1(I)$ et $\mathsf{L}^1(I)$. □

Remarque 4.5. Dans le cas où E est normé, la topologie de H^*, est celle induite par N_1. Donc $L^1_+(I)$ et $\ell^1(I)$ sont munis de N_1.

Si M_1, M_2 sont deux multimesures définies sur Ω à valeurs dans cc(E), dans [9], :

$$\sup(M_1, M_2)(A) = \sup\{M_1(B) + M_2(D) / \text{ (B, D) est une } \Omega\text{-partition de A}\},$$

la borne supérieure est prise dans cc(E). La semi-variation $M^{\cdot}$ de [9], est définie par :

$$\forall f \in H, M^{\cdot}(f) = \sup\{\|x\|, x \in \int |f||M|\}, \text{ où } |M| = \sup(M, -M).$$

Remarque 4.6. Si E est complet alors pour tout f de $\mathsf{L}^1(I)$, on a I(f)$\in$ cc(E). Si $M^{\cdot}$ désigne la semi-variation de [9], d'après l'inégalité $N_1 \leq M^{\cdot} \leq 2N_1$ sur H de [9] page 89 , on peut voir que $\ell^1(I)$ est l'espace $\ell^1(M^{\cdot})$ de [9] ou [10]. Cependant si E n'est pas complet, dans la méthode de [9], il se pose le problème du prolongement par continuité de l'intégrale qui se retrouve dans $cc(E'^*)$, où E'^* est le complété faible de E'. Ici, N_1 est définie à partir de I^*, ce qui permet de ne pas séparer l'intégrabilité d'avec le prolongement de l'intégrale, comme dans [9], et de ne pas perdre le lien avec l'espace E où tout prend ses valeurs.

Théorème 4.7. (α) *Les espaces $\pounds^1_+(I)$ et $\mathsf{L}^1(I)$ sont complets pour la topologie de la convergence en moyenne.*

(β) *Si $(f_n)_{n\in\mathbb{N}^*}$ est une suite de $\pounds^1_+(I)$ (resp. $\mathsf{L}^1(I)$) qui converge en moyenne vers f, alors $(f_n)_{n\in\mathbb{N}^*}$ possède une sous-suite qui converge I-pp vers f.*

Preuve. Preuve de (α) : Soit $(f_n)_{n\in\mathbb{N}^*}$ une suite de cauchy de $\pounds^1_+(I)$, il existe une sous-suite telle que :

$$N_1(f_{n_{k+1}} - f_{n_k}) < 2^{-k}; k = 1, 2, ... \tag{4.1}$$

D'après les propositions 3.4 (iii) et 3.8, on peut supposer les f_{n_k} définies partout. On obtient alors : $\forall y \in E', \delta^*(y/\int |f_{n_{k+1}} - f_{n_k}|I) = \int |f_{n_{k+1}} - f_{n_k}|I_y \leq 2^{-k}\|y\|$, $I_y[\sum_{k=1}^{k=p} |f_{n_{k+1}} - f_{n_k}|] \leq \sum_{k=1}^{k=+\infty} I_y[|f_{n_{k+1}} - f_{n_k}|] < +\infty$. D'après le théorème d'intégration termes à termes d'une série, $g = \sum_{k=1}^{k=\infty}[|f_{n_{k+1}} - f_{n_k}|]$ est I_y-intégrable pour tout y$\in$E'. D'après la proposition 2.4 (3), g est I-intégrable et donc g est finie I-pp. Posons : $f(t) = \sum_{k=1}^{k=+\infty}(f_{n_{k+1}} - f_{n_k})(t)$ *si* $g(t) < +\infty$, *et* $f(t) = 0$ *si* $g(t) = +\infty$. On a $f_{n_k} = f_{n_1} + (f_{n_2} - f_{n_1}) + ... + (f_{n_k} - f_{n_{k-1}})$. Donc $(f_{n_k})_{k\in\mathbb{N}^*}$ converge simplement vers f_{n_1} + f I-pp. De plus $0 \leq f_{n_k} \leq f_{n_1} + g$ I-pp, et $f_{n_1} + |g| \in L^1_+(I)$, donc, d'après le théorème 3.11, on a $(f_{n_1} + f) \in L^1_+(I)$, *et* $\lim_{n_k\to+\infty} \int f_{n_k} I = \int (f_{n_1} + f)I$ au sens de l'ordre. Il reste à montrer que $(f_n)_{n\in\mathbb{N}^*}$ converge en moyenne vers h = f_{n_1} + f. Soient $g_k = f_{n_k}$, $k \geq 1$, et $\Phi_{n_k} = \inf\{g_i, n \leq i \leq n+k\} \in \pounds^1_+(I), n \geq 1, k \geq 0$. On a $\Phi_{n_k} \in L^1_+(I)$, et donc, d'après le théorème 3.11, $\Phi_n = \inf_{k\geq 0} \Phi_{n_k} = \inf_{k\geq n} g_k \in \pounds^1_+(I)$. D'après 4.1, on a :

$$N_1(g_{n+1} - g_n) < 2^{-n}, \forall n \in \mathbb{N}^*. \tag{4.2}$$

Pour tout t, il existe i=i(t), $0 \leq i \leq k$ tel que Φ_{n_k}(t) = g_{n+i}(t). Si $i \geq 1$ on a : $0 \leq g_n(t) - \Phi_{n_k}(t)$ et, $g_n(t) - \Phi_{n_k}(t) = g_n(t) - g_{n+i}(t) = (g_n(t) - g_{n+1}(t)) + ... + (g_{n+i-1}(t) - g_{n+i}(t))$. On en déduit :

$$0 \leq |g_n(t) - \Phi_{n_k}(t)| \leq |g_n(t) - g_{n+i}(t)| \leq |(g_n(t) - g_{n+1}(t))| + ... + |(g_{n+i-1}(t) - g_{n+i}(t))|.$$

Ceci est vrai aussi si i=0 donc : $0 \leq |g_n - \Phi_{n_k}| \leq |g_n - g_{n+i}| \leq |(g_n - g_{n+1})| + \ldots + |(g_{n+i-1} - g_{n+i})|$. D'après 4.2, $N_1(g_n - \Phi_{n_k}) \leq 2^{-n} + \ldots + 2^{-n-k+1} \Rightarrow N_1(g_n - \Phi_{n_k}) \leq 2^{-n+1}$, d'où $\delta^*(y/I(|g_n - \Phi_{n_k}|)) \leq 2^{-n+1}\|y\|$. Or $(g_n - \Phi_{n_k}) \uparrow (g_n - \Phi_n) \in \pounds^1(I)_+$, donc d'après le théorème 3.11, $I(g_n - \Phi_n) = \sup_{k \geq 0} I(g_n - \Phi_{n_k})$. On en déduit:

$$\delta^*\left(y/I(|g_n - \Phi_n|)\right) = \sup_{k \geq 0} \delta^*\left(y/I(|g_n - \Phi_{n_k}|)\right) \leq 2^{-n+1}\|y\|, \text{ et}$$

$$N_1(g_n - \Phi_n) \leq 2^{-n+1} \; \forall n \geq 1. \tag{4.3}$$

Prenons $\Psi_{n_k} = \sup\{g_i,\ n \leq i \leq n+k\}$, $\Psi_n = \sup_{k \geq 0} \Psi_{n_k} = \sup_{k \geq n} g_k$.

On a $\Psi_{n_k} \uparrow \Psi_n$ et $\Psi_n \leq \sup_{k \geq 1} g_k \leq g_1 + g$ I-pp, donc $\Psi_n \in L^1_+(I)$. En raisonnant comme pour les Φ_{n_k}, on a : $N_1(\Psi_{n_k} - g_n) \leq 2^{-n+1}$, $(\Psi_{n_k} - g_n) \uparrow (\Psi_n - g_n)$. On obtient donc :

$$N_1(\Psi_n - g_n)) \leq 2^{-n+1} \; \forall n \geq 1. \tag{4.4}$$

Nous avons aussi $\Phi_n \leq h \leq \Psi_n$,$\Phi_n \uparrow$, et $\Psi_n \downarrow$. On a $|\Phi_n - h| \leq |\Phi_n - \Psi_n|$, et $|\Psi_n - h| \leq |\Phi_n - \Psi_n|$, or $N_1(\Psi_n - \Phi_n) \leq N_1(\Psi_n - g_n) + N_1(\Phi_n - g_n))$, si on utilise 4.3 et 4.4, on a :

$$N_1(\Psi_n - \Phi_n) \leq 2^{-n+2},\ N_1(\Phi_n - h) \leq 2^{-n+2} \text{ et } N_1(\Psi_n - h) \leq 2^{-n+2}. \tag{4.5}$$

De plus pour $n \geq 1$, on a $h - \Psi_n \leq h - g_n \leq h - \Phi_n$, car $\Phi_n \leq g_n \leq \Psi_n$. Cela donne : $|h - g_n| \leq |h - \Psi_n| + |h - \Phi_n| = \Psi_n - h + h - \Phi_n = \Psi_n - \Phi_n$ donc, d'après 4.5, $N_1(h - g_n) \leq 2^{-n+2}$. On déduit : $\lim_{n \to +\infty} N_1(g_n - h) = 0 = \lim_{n \to +\infty} N_1(\Phi_n - h) = \lim_{n \to +\infty} N_1(\Psi_n - h)$ et donc :

$$\lim_{n_k \to +\infty} N_1(f_{n_k} - h) = 0 \text{ et } N_1(f_n - h) \leq N_1(f_{n_k} - f_n) + N_1(f_{n_k} - h).$$

Pour $n \leq n_k$ on a $\lim_{n_k \to +\infty} N_1(f_{n_k} - f_n)$=0 ie f_n converge en moyenne vers h, et donc $\pounds^1_+(I)$ est complet. On a $\pounds^1_+(I)$ est complet donc fermé dans F(T,$\overline{\mathbb{R}}$). Comme $H_+ \subset \pounds^1(I)$, on a $\mathsf{L}^1_+(I) = \overline{H}_+ \subset \pounds^1_+(I)$, et $\mathsf{L}^1_+(I)$ est complet. Si $(f_n)_{n \in \mathbb{N}^*}$ est une suite de cauchy de $\mathsf{L}^1(I)$ alors $(f_n^+)_{n \in \mathbb{N}^*}$, $(f_n^-)_{n \in \mathbb{N}^*}$ sont de Cauchy, car, $|f_n^+ - f_m^+)| \leq |f_n - f_m|$, et $|f_n^- - f_m^-| \leq |f_n - f_m|$. Si $f^+ = \lim_{n \to +\infty} f_n^+$ et $f^- = \lim_{n \to +\infty} f_n^-$ dans $\mathsf{L}^1_+(I)$, on a (f_n) converge en moyenne vers $f = f^+ - f^-$ car $|f_n - f| \leq |f_n^+ - f^+| + |f_n^- - f^-|$ I-pp, et donc $\mathsf{L}^1(I)$ est complet. Preuve de (β) : Supposons que $(f_n)_{n \geq 0}$ converge en moyenne vers f. Alors $(f_n)_{n \geq 0}$ est de Cauchy et par suite, la preuve de (α) montre l'existence d'une sous-suite $(f_{n_k})_{k \geq 0}$ convergente dont la limite ne peut être que f. □

On a : $|f_n^+ - f^+)| \leq |f_n - f|$ et $|f_n^- - f^-)| \leq |f_n - f|$. Avec la proposition 2.8, on obtient que $I^*(f_n^+)$ et $I^*(f_n^-)$ convergent respectivement vers $I^*(f^+)$ et $I^*(f^-)$ dans cfb(E). De plus on a, si $y \in E'$: $\delta^*(y/\int f_n I) = \delta^*(y/\int f_n^+ I) + \delta^*(y/\int(-f_n^-)I)$, de même pour f. On en déduit : $|\delta^*(y/\int f_n I) - \delta^*(y/\int f I)| \leq$

$$|\delta^*(y/\int f_n^+ I) - \delta^*(y/\int f^+ I)| + |\delta^*(y/-\int f_n^- I) - \delta^*(y/-\int f^- I)|.$$

Ainsi $DH[I(f_n),I(f)] \leq DH[I(f_n^+),I(f^+)] + DH[-I(f_n^-),-I(f^-)]$, qui tend vers o. On en déduit la remarque suivante.

Remarque 4.8. Supposons que $(f_n)_{n\geq 0}$ converge vers f dans $\mathsf{L}^1(I)$. Alors d'après la preuve du théorème 4.7, $(I(f_n))_{n\geq 0}$ converge vers I(f) pour la topologie de Hausdorff de cfb(E).

Proposition 4.9. *On a :* $\mathsf{L}^1(I) \subset \pounds^1_+(I)$*, et* $\ell^1_+(I) \subset L^1_+(I) \cap \mathbb{R}^T$.

Preuve. Les espaces $\pounds^1_+(I)$, et $L^1_+(I) \cap \mathbb{R}^T$ sont complets, donc, respectivement, fermés dans $F_+(T,\overline{\mathbb{R}})$ et H^*. On a $H_+ \subset \pounds^1_+(I)$, et $H_+ \subset L^1_+(I) \cap \mathbb{R}^T$. En prenant les adhérences dans $F(T,\overline{\mathbb{R}})$, puis dans H^*, on obtient : $\mathsf{L}^1(I) \subset \pounds^1_+(I)$, et, $\ell^1_+(I) \subset L^1_+(I) \cap \mathbb{R}^T$. □

$$NOTONS : [\Delta]_C = \{\, f \in \Delta, I(f) \in cc(E)\}$$

Remarque 4.10. D'après 2) et 3) de la proposition 4.9, on a : $[\ell^1_+(I)]_C \subset [L^1_+(I)]_C \cap \mathbb{R}^T \subset [L^1_+(I)]_C \subset L^1_+(I) \subset \pounds^1_+(I)$. L'espace $[\ell^1(I)]_C$ correspond à celui obtenu par [12] pour l'intégrale multivoque, et $[L^1_+(I)]_C \cap \mathbb{R}^T$ pour l'intégrale de Daniell secondaire. L'espace $L^1_+(I)$ est celui que nous avons obtenus dans [5] ou [7] pour l'intégrale de Daniell secondaire, et nous venons de l'agrandir par $\pounds^1_+(I)$ en y mettant une topologie, ce qui n'existait pas dans [12] et [5] où les théorèmes de convergence étaient donnés au sens de l'ordre. Ici nous obtenons le théorème 4.17 ci après. L'espace $\mathsf{L}^1(I)$ est celui que nous venons d'obtenir pour l'intégrale multivoque. L'espace $\ell^1(I)$ est celui obtenu par [9], si E est un espace de Banach. D'autre part [12] a défini, dans le cas cfb(E), une fonction intégrable comme une fonction positive limite croissante de fonctions de H, et dont l'intégrale supérieure est dans cfb(E). Une telle fonction est intégrable pour notre définition, d'après le théorème de la convergence monotone. Nous venons donc d'unifier trois méthodes d'intégrations de [12], en les généralisant à cf(E), et le lien avec la méthode de [9] est maintenant établi.

D'après la remarque 2.9, $[\ell^1(I)]_C = \overline{H}$ dans $[H^*]_C$. De plus, si E est complet alors, cc(E) est fermé dans cfb(E). On en déduit la proposition suivante.

Proposition 4.11. *On a* $[\ell^1(I)]_C = \overline{H}$ *dans* $[H^*]_C$*. Si E est complet, alors :* $[\ell^1(I)]_C = \ell^1(I)$*, et* $[\mathsf{L}^1(I)]_C = \mathsf{L}^1(I)$*, où* $[\mathsf{L}^1(I)]_C = \overline{H}$ *dans* $[F(T,\overline{\mathbb{R}})]_C$.

Proposition 4.12. *Les espaces quotient de* $\mathsf{L}^1(I)$ *et* $\ell^1(I)$ *sont des espaces de Banach.*

Cela résulte du théorème 4.7. □

Nous rappelons dans l'ordre, les théorèmes 4.1, 4.2, et 4.3 de [12] pages 132 à 135. Ces théorèmes n'existait dans [12], que pour l'intégrale multivoque : pour l'intégrale de Daniell secondaire, comme dans [5], il n'y avait pas de topologie, et les théorèmes était au sens de l'ordre.

Théorème 4.13. *Si* $(K_n)_{n\geq 1}$ *est une suite d'éléments de cc(E), telle que* $0 \in K_n$ *pour tout* $n \in \mathbb{N}^*$*. Si* $K = \sup_n \sum_{i=1}^{i=n} K_i \in cc(E)$*, alors K est la limite, au sens de Hausdorff dans cc(E), de la suite* $\sum_{i=1}^{i=n} K_i$.

Dans toute la suite de cette section, on notera : $\sup_n \sum_{i=1}^{i=n} K_i$ la limite au sens de l'ordre de la suite $\sum_{i=1}^{i=n} K_i$; et par $\sum_{i\geq 1} K_i$, la limite pour la topologie de Hausdorff de la suite $\sum_{i=1}^{i=n} K_i$.

Théorème 4.14.

(1) Si $(f_n)_{n\in\mathbb{N}^*}$ est une suite croissante de $[\ell^1_+(I)]_C$, si $f = \sup_{n\in\mathbb{N}^*} f_n$ et s'il existe $g \in \mathbb{R}^T$ telle que $f_n \leq g$ pour tout $n \in \mathbb{N}^*$, et si $I^*(g) \in cc(E)$ alors $f \in [\ell^1_+(I)]_C$ et $\lim_{n\to+\infty} \int f_n I = \int fI$ pour la topologie de Hausdorff de cc(E).

(2) Si $(f_n)_{n\in\mathbb{N}^*}$ est une suite décroissante de $[\ell^1_+(I)]_C$, alors : $\lim_{n\to+\infty} \int f_n I = \int fI$ pour la topologie de Hausdorff de cc(E).

Théorème 4.15. *Si $(f_n)_{n\in\mathbb{N}^*}$ est une suite de $[\ell^1(I)]_C$, qui converge simplement vers une fonction f, s'il existe $g \in \mathbb{R}^T$ telle que $|f_n| \leq g$ pour tout $n \in \mathbb{N}^*$, et si $I^*(g) \in cc(E)$, alors $f \in [\ell^1(I)]_C$, et, $\lim_{n\to+\infty} \int f_n I = \int fI$ pour la topologie de Hausdorff de cc(E).*

Remarque 4.16. Le théorème 4.17 suivant, ainsi que son corollaire 4.19 sont des versions I-pp, des deux théorèmes, ci-avant énoncés de [12]. La notion de I-négligeable que nous avons introduit, nous a permis d'énoncer ces théorèmes, pour des fonctions définies I-pp sur T, à valeurs dans $\overline{\mathbb{R}}$. Nous obtenons également que, ces théorèmes de convergence, sont des théorèmes de convergence en moyenne.

Théorème 4.17. *(1) Si $(f_n)_{n\in\mathbb{N}^*}$ est une suite croissante I-pp de $£^1_+(I)$, si $f = \sup_{n\in\mathbb{N}^*} f_n$ et S'il existe $g \in F(T,\overline{\mathbb{R}})$ telle que $f_n \leq g$ I-pp pour tout $n \in \mathbb{N}^*$, et si $I^*(g) \in cc(E)$ alors $f \in [£^1_+(I)]_C$ et $(f_n)_{n\in\mathbb{N}^*}$ converge en moyenne vers f.*

(2) Si $(f_n)_{n\in\mathbb{N}^*}$ est une suite décroissante I-pp de $[£^1_+(I)]_C$, et si $f = \inf_{n\in\mathbb{N}^*} f_n$, alors : $f \in [£^1_+(I)]_C$ et $(f_n)_{n\in\mathbb{N}^*}$ converge en moyenne vers f.

(3) Si $(f_n)_{n\in\mathbb{N}^*}$ est une suite de $[£^1_+(I)]_C$, qui converge simplement I-pp vers une fonction f, s'il existe $g \in F(T,\overline{\mathbb{R}})$ telle que $|f_n| \leq g$ I-pp pour tout $n \in \mathbb{N}^*$, et si $I^*(g) \in cc(E)$, alors $f \in [£^1_+(I)]_C$ et $(f_n)_{n\in\mathbb{N}^*}$ converge en moyenne vers f. Dans les trois cas on a $\lim_{n\to+\infty} \int f_n I = \int fI$ pour la topologie de cc(E).

Preuve. La preuve est la même que celle du théorème 4.2 page 133 pour l'intégrale multivoque de [12], en y mettant la notion I-pp puis en utilisant le lemme 4.18 ci-après. Cette preuve est basée sur le théorème 4.1 [12] (cf théorème 4.13), qui permet de ramener une convergence au sens de l'ordre, à celle pour la topologie de Hausdorff de cc(E). Preuve de (1) : Une fonction intégrable est finie I-pp. Si nous posons alors : $\Phi_i = f_i - f_{i-1}$ I-pp, $\forall i \geq 1$ avec $f_0 = 0$. On a :

$$f_n = \Sigma_{i=1}^{i=n} \Phi_i \;\; I-pp. \tag{4.6}$$

On a aussi $f - f_n = \sup_{k>n} \Sigma_{i=n+1}^{i=k} \Phi_i = \Sigma_{i>n} \Phi_i \; I-pp$. Les fonctions Φ_i sont I-intégrables. Ainsi $I^*(\Phi_i) = I(\Phi)$, et $I^*(f_n) = I(f_n)$. Donc, d'après la proposition 2.4 (4) on a

$$I^*(f - f_n) = I^*\left(\sup_{k>n}\left(\Sigma_{i=n+1}^{i=k} \Phi_i\right)\right) = \sup_{k>n} I\left(\Sigma_{i=n+1}^{i=k} \Phi_i\right) = \sup_{k>n} \Sigma_{i=n+1}^{i=k} I(\Phi_i).$$

De plus la multimesure étant monotone, on a : $0 \in I(\Phi_i))$, $\forall$ i. Comme l'inégalitée $f - f_n \leq f \leq g$ I-pp, entraîne $I^*(f - f_n) \in cc(E)$, donc, on peut appliquer le théorème 4.1 de [12] à la suite de terme général $I^*(\Phi_i) = I(\Phi_i)$, On a : $I^*(f - f_n) = \lim_{k \to +\infty} \Sigma_{i=n+1}^{i=k} I^*(\Phi_i) = \Sigma_{i>n} I^*(\Phi_i)$, la limite étant prise au sens de Hausdorff. D'après 4.6, on a $I(f_n) = I\left(\Sigma_{i=1}^{i=n} \Phi_i\right)$, de plus $\sup_{n \in \mathbb{N}^*} (f_n) \leq g$ I-pp, et $I^*(g) \in cc(E)$. On en déduit que : $\sup_{n \in \mathbb{N}^*} I(f_n) = \sup_{n \in \mathbb{N}^*} \Sigma_{i=1}^n I(\Phi_i) \in cc(E)$, avec le théorème 4.1 [12], si les limites sont au sens de Hausdorff dans cc(E), on a :

$$\sup_{n \in \mathbb{N}^*} \Sigma_{i=1}^n I(\Phi_i) = \lim_{n \to +\infty} \Sigma_{i=1}^{i=n} I(\Phi_i) = \lim_{n \to +\infty} I(f_n)$$

et donc $\lim_{n \to +\infty} I^*(f - f_n) = \Sigma_{i>n} I^*(\Phi_i) = \{0\}$.
Comme $\{0\} = \lim_{n \to +\infty} I^*(f - f_n) = \lim_{n \to +\infty} DH(\{0\}, I^*(f - f_n)) = \lim_{n \to +\infty} N_1(f - f_n)$, on a : $(f_n)_{n \in \mathbb{N}^*}$ converge en moyenne vers f. D'après le lemme 4.18 on a $\lim_{n \to +\infty} I(f_n) = \lim_{n \to +\infty} I(f)$, pour la topologie de Hausdorff de cc(E), d'où (1). Preuve de (2) : On suppose maintenant que $f_n \downarrow f$ I-pp. On a : $(f_1 - f_n) \uparrow (f_1 - f)$ I-pp, et $(f_1 - f_n) \leq f_1$ I-pp, avec , $I(f_1) \in cc(E)$. Donc d'après ce qui précède $(f_1 - f_n)_{n \in \mathbb{N}^*}$ converge en moyenne vers $(f_1 - f)$. Donc d'après le lemme 4.18, $\lim_{n \to +\infty} I(f_1 - f_n) = \lim_{n \to +\infty} I(f_1 - f)$, pour la topologie de Hausdorff de cc(E). De plus pour tout $n \in \mathbb{N}^*$, on a : $I(f_1) = \int f_1 I = \int f_n I + \int (f_1 - f_n) I$, et $\int f_1 I = \int f I + \int (f_1 - f) I$, on en déduit que :

$$\delta^*\left(y / \int (f_1 - f) I\right) - \delta^*\left(y / \int (f_1 - f_n) I\right) = \delta^*\left(y / \int f_n I\right) - \delta^*\left(y / \int f I\right).$$

Avec la formule d'Hörmander, on a : $DH(I(f_1 - f), I(f_1 - f_n)) = DH(I(f_n), I(f))$. Vu que $0 = \lim_{n \to +\infty} DH(I(f_1 - f), I(f_1 - f_n)) = \lim_{n \to +\infty} DH(I(f_n), I(f))$, on a $I(f_n) = \int f_n I$ converge vers $I(f) = \int f I$ au sens de Hausdorff dans cc(E). Du lemme 4.18, on déduit que (f_n) converge en moyenne vers f, d'où (2). Preuve de (3): Comme dans la preuve du théorème 4.7, on considère les fonctions : $\Phi_n = \inf_{p \geq n} f_p$ et $\Psi_n = \sup_{p \geq n} f_p$. On a, par hypothèse : $0 \leq \Phi_n \leq \Psi_n \leq g$, et $I^*(g) \in cc(E)$. Donc d'après l'assertion (3) du théorème 3.11, on a Φ_n, $\Psi_n \in [\pounds_+^1(I)]_C$. Avec (1) et (2), on a Φ_n et Ψ_n convergent en moyenne vers f. De plus nous avons : $\Phi_n \leq f \leq \Psi_n \Rightarrow |f_n - f| \leq |\Phi_n - f| + |\Psi_n - f|$, on en déduit que $N_1(f_n - f) \leq N_1(\Phi_n - f) + N_1(\Psi_n - f)$, et donc (f_n) converge en moyenne vers f. Ainsi, vu que $DH(I(f_n), I(f)) \leq N_1(f_n - f)$, $I(f_n)$ converge vers I(f) pour la topologie de Hausdorff de cc(E). □

Lemme 4.18. *(1) Si $(f_n)_{n \in \mathbb{N}^*}$ est une suite de $\pounds_+^1(I)$, si $f \in \pounds_+^1(I)$ et si $f_n \leq f$ pour tout $n \in \mathbb{N}^*$, on a : $[I(f - f_n) \to 0$ dans $cfb(E)] \Leftrightarrow [I(f_n) \to I(f)$, dans cfb(E)$] \Leftrightarrow [f_n \to f$ en moyenne$]$.*

(2) De même si $f_n \geq f$ pour tout $n \in \mathbb{N}^*$ on a : $[I(f_n - f) \to 0$ dans cfb(E)$] \Leftrightarrow [I(f_n) \to I(f)$, dans cfb(E)$] \Leftrightarrow [f_n \to f$ en moyenne$]$.

Preuve. (1) : On a $f = f - f_n + f_n$ donc $I(f) = I(f - f_n) + I(f_n)$ et on a

$$\delta^*(y / I(f - f_n)) = \delta^*(y / I(f)) - \delta^*(y / I(f_n)) \geq 0, \text{ i.e.}$$

$$N_1(f-f_n)=\sup\{|\delta^*(y/I(f))-\delta^*(y/I(f_n))|, y\in E', \|y\|\le 1\},$$

et donc $N_1(f-f_n)=DH(I(f_n),I(f))$, d'où le résultat. Pour le (2) il suffit, dans la preuve du (1), d' intervertir f_n et f, car $f_n \ge f$. □

Corollaire 4.19. *Si $(f_n)_{n\in\mathbb{N}^*}$ est une suite de $\mathsf{L}^1(I)$, qui converge I-pp vers f, s'il existe $g\in\mathsf{L}^1_+(I)$ telle que $\int gI\in cc(E)$ et $|f_n|\le g$ I-pp pour tout $n\in\mathbb{N}^*$, alors $f\in\mathsf{L}^1(I)$ et $\int fI = \lim_{n\to+\infty}\int f_nI$ au sens de la topologie de Hausdorff de cc(E), et $(f_n)_{n\in\mathbb{N}^*}$ converge en moyenne vers f.*

Preuve. On applique le théorème 4.17 (3), aux suites $(f_n^+)_{n\in\mathbb{N}^*}$ et $(f_n^-)_{n\in\mathbb{N}^*}$ et l'on en déduit que : $(f_n^+)_{n\in\mathbb{N}^*}$ converge en moyenne vers f^+, $(f_n^-)_{n\in\mathbb{N}^*}$ converge en moyenne vers f^-, on a aussi $f^+,f^-\in\pounds^1_+(I)$, et

$$\lim_{n\to+\infty}\int f_n^+I=\int f^+I,\ \lim_{n\to+\infty}\int f_n^-I=\int f^-I, \tag{4.7}$$

au sens de la topologie de Hausdorff de cc(E). Vu que $f_n^+,f_n^-\in\mathsf{L}^1_+(I)$, et $\mathsf{L}^1_+(I)$ est complet, donc fermé dans $\pounds^1_+(I)$, on a : $f=f^+-f^-\in\mathsf{L}^1_+(I)-\mathsf{L}^1_+(I)=\mathsf{L}^1(I)$. Comme dans la remarque 4.8, on a :
$|\delta^*(y/\int f_nI)-\delta^*(y/\int fI)|\le$

$$|\delta^*(y/\int f_n^+I)-\delta^*(y/\int f^+I)|+|\delta^*(y/-\int f_n^-I)-\delta^*(y/-\int f^-I)|.$$

On en déduit que : $DH(I(f_n),I(f))=DH(I(f_n^+),I(f^+))+DH(-I(f_n^-),-I(f^-))$. Avec 4.7, on a $\int fI=\lim_{n\to+\infty}\int f_nI$ au sens de la topologie de Hausdorff de cc(E). Ce résultat, appliqué à la suite $(f_n-f)_{n\in\mathbb{N}^*}$ montre que f_n converge en moyenne vers f. □

Remarque 4.20. (1) Nous n'avons pas supposé que E est un espace de Banach, ni qu'il possède la propriétée de Bessaga-Pelszyǹsky, voir définition V.15 page 75 [9].

(2) Si E est un Banach et si M est à valeurs dans cfb(E), alors M est à valeurs faiblement compactes, d'après le théorème 2 de [6] ou I-2-2 [7]. D'après la proposition 2.2 3), si $f\in[\pounds^1_+(I)]_C$ alors f est faiblement intégrable (voir définition VI.31 page 96 [9], et le théorème VI.6 page 97 montre que cela est équivalent à dire que f est dans l'espace de Pallu De La Barrière $L^1(M^\cdot)$, et donc que les espaces coïncident.

(3) Dans la section 5 nous donnerons des exemples de fonctions dont l'intégrale est à valeurs dans cfb(E), et non dans cc(E).

Théorème 4.21. *Si $(f_n)_{n\in\mathbb{N}^*}$ est une suite de $\pounds^1_+(I)$ telle que, $\sup_{n\in\mathbb{N}^*}\Sigma_{k=1}^{k=n}\int f_kI\in$cfb(E), alors $\Sigma_{n\in\mathbb{N}^*}f_n\in\pounds^1_+(I)$ et $\int(\Sigma_{n\in\mathbb{N}^*}f_n)I=\sup_{n\in\mathbb{N}^*}\Sigma_{k=1}^{k=n}\int f_kI.$*

Preuve. En appliquant le théorème 3.11 à la suite croissante $(u_n)_{n\in\mathbb{N}^*}$, où $u_n=\Sigma_{k=1}^{k=n}f_k\in\pounds^1_+(I)$, on obtient le résultat. □

Corollaire 4.22. (Lemme de Borel-Cantelli) *Si* $(A_n)_{n\in\mathbb{N}^*}$ *est une famille de parties de T, telle que* $1_{A_n} \in L^1_+(I)$, $\forall n \in \mathbb{N}^*$, *et si* $\sup_{n\in\mathbb{N}^*} \Sigma_{k=1}^{k=n} \int 1_{A_k} I \in cfb(E)$*, Alors : sauf sur un ensemble I-négligeable, t n'appartient qu'à un nombre fini de* A_n.

En effet, si on applique le théorème 4.21 à la suite $(u_n)_{n\in\mathbb{N}^*}$, avec $u_n = \Sigma_{k=1}^{k=n} 1_{A_k} \in L^1_+(I)$, on a $\sup_{n\in\mathbb{N}^*} \Sigma_{k=1}^{k=n} 1_{A_k} \in L^1_+(I)$, et donc est fini I-pp. □

Théorème 4.23. (Théorème de Lebesgue pour les séries) *Si* $(f_n)_{n\in\mathbb{N}^*}$ *est une suite de* $L^1(I)$*, telle que* $\Sigma_{n\in\mathbb{N}^*} \int |f_n| I \in cc(E)$*, alors* $\Sigma_{n\in\mathbb{N}^*} f_n \in L^1(I)$ *et* $\int (\Sigma_{n\in\mathbb{N}^*} f_n) I = \Sigma_{n\in\mathbb{N}^*} \int f_n I$.

Il suffit d'appliquer le corollaire 4.19 à la suite $u_n = \Sigma_{k=1}^{k=n} f_k$. La convergence ici est au sens de la topologie de Hausdorff de cc(E) : voir notations après le théorème 4.13. □

5 Exemples de Fonctions Integrables dont L'integrale n'Est pas dans cc(E)

Soit $(\mathbb{R}, \mathbb{B}(\mathbb{R}), m)$ où m est une mesure finie sur $\mathbb{R}$ admettant une densitée par rapport à la mesure de Lebesgue. On désigne par $\mathbb{B}(\mathbb{R})$, la tribu borélienne de $\mathbb{R}$, et on pose, pour tout $A \in \mathbb{B}(\mathbb{R}) : N(A) = \{f \in L^1(m), 0 \leq f \leq 1_A\}$. Posons, pour $n \in \mathbb{N}^*, A_n = [n, n+1[$ et

$$E = \{\, f \in L^1(m), f \text{ est nulle sur les } A_n \text{ en dehors d'un nombre fini}\} \;.$$

E est un sous-espace de Riesz de $L^1(m)$. On a N(A)$\in$ cc($L^1(m)$). Nous supposons aussi m(A_n) non nul pour tout $n \in \mathbb{N}^*$. Soit M : $\mathbb{B}(\mathbb{R})\longrightarrow$cfb(E) avec M(A) = N(A)$\cap$E. D'après le théorème de décomposition de Riesz, M est une multimesure faible monotone. De plus M(A_n)= N(A_n), on obtient donc que $M(A_n) \in cc(E)$. De plus, pour tout $n \in \mathbb{N}^*$ tel que $n \geq 2$, on a [-n,n[$=\cup_{k=-n}^{k=n-1}[p, p+1[$. On obtient alors que M([-n,n[)$\in$cc(E) et $1_{[-n,n[} \uparrow 1_{\mathbb{R}}$. Nous avons donc que M est une multimesure faible monotone s-compacte. Soit A $\in \mathbb{B}(\mathbb{R})$ tel que m($A \cap A_n$)> 0 pour une infinité de n, par exemple $A = \mathbb{R}$ ou tout interval non borné de $\mathbb{R}$. Ainsi, on a : $1_A \notin E$, et donc $1_A \notin M(A)$. Si $f_n = \Sigma_{p=-n}^{p=n} 1_{A\cap A_p}$ alors $(f_n)_{n\in\mathbb{N}^*}$ converge simplement vers $f = \Sigma_{p\in\mathbb{Z}} 1_{A\cap A_p} = 1_A$. Vu que $0 \leq f_n \leq 1_A$ et que $1_A \in L^1(m)$, d'après le théorème de convergence dominée dans $L^1(m)$, on a $(f_n)_{n\in\mathbb{N}^*}$ converge vers 1_A dans $L^1(m)$. De plus $f_n \in M(A)$, et la limite de $(f_n)_{n\in\mathbb{N}^*}$, i.e. $1_A \notin M(A)$, donc $M(A) \notin cc(E)$. Nous avons aussi : $M(A) = \int 1_A M \in cfb(E)$, i.e. $1_A \in L^1(M) \cap \mathbb{R}^T$, et pourtant $M(A) = \int 1_A M \notin cc(E)$. Ainsi 1_A n'est pas intégrable au sens de [12]. On a aussi que E n'est pas complet car non fermé dans $L^1(m)$.

Remarque 5.1. Nous obtenons ainsi des fonctions intégrables dans notre sens et qui ne le sont pas pour [12]. Nous avons aussi obtenu que si E n'est pas complet, une multimesure faible monotone s-compacte à valeurs cfb(E) peut ne pas être à valeurs cc(E). Cependant, d'après le théorème 2 de [6], si E est complet, une telle multimesure est à valeurs cc(E).

Proposition 5.2. *Si E n'est pas complet, cc(E) n'est plus nécessairement fermé dans cfb(E), ni complet.*

Preuve. Vu que E est dense dans $L^1(m)$, on peut prendre $E' = L^\infty(m)$. Si on désigne par N_1 la semi-norme associée à la multimesure N, et M_1 celle pour M. Soit $B \in \mathbb{B}(\mathbb{R})$, on a pour tout $g \in L^\infty(m)$, $\delta^*(g, N(B)) = \int 1_B g^+ m \geq \delta^*(g, M(B))$, et donc : $M_1(1_B) \leq N_1(1_B) = \sup\{\int 1_B g^+ m, g \in L^\infty(m), \|g\|_\infty \leq 1\} \leq \int 1_B m$. Soient A et f_n comme ci avant, on a $M_1(1_A - f_n) \leq \int (1_A - f_n)m$. Nous avons alors que f_n converge en moyenne vers 1_A dans ℓ^1(M), et donc $\int f_n M$ converge vers $\int 1_A M$ pour la topologie de Hausdorff, $\int f_n$M$\in$cc(E), et pourtant $\int 1_A$M$\notin$cc(E). □

Remarque 5.3. La preuve précédente montre que $1_A \in \ell^1(M)$ et $1_A \notin [\ell^1(M)]_C$. Donc l'espace $[\ell^1(M)]_C$ de [12] est strictement contenu dans notre ℓ^1(M). Le problème est de savoir si l'inclusion ℓ^1_+(M)$\subset$ L^1_+(M) ou L^1_+(M) $\subset$ $\pounds^1_+$(M) peut être stricte. En d'autres termes une fonction intégrable pour l'intégrale de Daniell secondaire l'est elle pour l'intégrale multivoque? ou encore H_+ est il dense dans $\pounds^1_+$(M)? Si E est complet la réponse est oui car nous revenons alors au cas compact comme mentionné dans la remarque 4.20, voir cependant la remarque 4.6.

6 Integration par Rapport a une Mesure Vectorielle Relativement Faiblement Compacte

L'objectif de cette section est le théorème 6.5, ci après, qui généralise les théorèmes 5.1 et 5.2 pages 144 et 145 de [12]. On applique une méthode de [12]. On suppose que E est un espace de Banach. Soit $m : \Omega \longrightarrow E$ une mesure vectorielle ie vérifiant : m($\emptyset$)= 0 et si $\{A_n$,n= 1,2 ...$\}$ est une Ω-partition de A $\in \Omega$, alors $m(A) = \lim_{n\to+\infty} \Sigma_{i=1}^{i=n} m(A_i)$ pour la topologie de E. m est relativement faiblement compacte si : {m(B),B$\in$ Ω,B $\subset A$ } est relativement faiblement compacte pour tout A $\in \Omega$. Comme dans [12] page 140, voir aussi [6] ou [7], Pour tout $A \in \Omega$, on définit $M_m(A)$, comme l'adhérence dans E, de toutes les sommes $\Sigma_{i\in I}\lambda_i m(A_i)$, où $\{A_i, i \in$ I$\}$ est une Ω-partition de A, et où $\lambda_i \in \mathbb{R}$ avec $|\lambda_i| \leq 1$. Si $|m_y|$ est la variation totale de m_y=$\langle m(\cdot), y\rangle, y \in E'$, alors d'après lemme 5.1 page141 [12], on a : $\delta^*(y : M_m(A))$= $|m_y|$(A), et donc $M_m : \Omega \longrightarrow$cc(E) est une multimesure forte et monotone, i.e. : 0$\in M_m(A)$ pour tout A$\in \Omega$, M_m est additive et si $\{A_n$, n= 1,2 ...$\}$ est une Ω-partition de A$\in \Omega$, alors $M_m(A) = \lim_{n\to+\infty} \Sigma_{i=1}^{i=n} M_m(A_i)$, pour la topologie de cc(E). De plus M_m est s-compacte.

Définition 6.1. On pose $\mathsf{L}^1(m)$= $\mathsf{L}^1(M_m)$ et $\ell^1(m)$ =$\ell^1(M_m)$.

Si $h = \Sigma a_i 1_{A_i}$ est une fonction en escaliers, alors $h \in H$, soit : $\int hm = \Sigma a_i m(A_i)$. L'application $\int(.)m$ ainsi définie de H dans E est une selection de $\int(.)M_m$, i.e. : $\int hm \in \int hM_m, \forall h \in H$. De la continuitée de $\int(.)M_m$ de H muni de la topologie induite par $F(T, \overline{\mathbb{R}})$ (resp. H^*) dans E Banach, H étant dense dans $\mathsf{L}^1(m)$ (resp. $\ell^1(m)$), on prolonge $\int(.)m$ par continuité en une application encore notée $\int(.)m$ de $\mathsf{L}^1(I)$(resp. $\ell^1(m)$ dans E qui est additive et continue. Si $f \in \mathsf{L}^1(m)$ (resp. $\ell^1(m)$), la valeur en f du prolongement de $\int(.)m$ est l'intégrale de f et est notée $\int fm$. D'après le lemme 5-2 page 142 [12], d(M_m(A),0)=$\bar{m}(A)$ où $\bar{m}$ est la semi-variation de m et d la distance de Hausdorff. On a donc $N_1(1_A)$ =$\bar{m}(A)$$\leq$ $|m|(A)$, A$\in \Omega$ où N_1 est la semi-norme de la topologie de la convergence en moyenne et $|m|$ est la variation de m. Soit $N_2(f) = \int |f||m|$ pour $f \in H$, si on prolonge

N_2 à $\overline{\mathbb{R}}_+^T$, comme dans [9], voir aussi [11], on a : si $f \in \overline{\mathbb{R}}_+^T$ alors $N_1(f) \leq N_2(f) = |m|^*(f)$. Si $f \in \overline{\mathbb{R}}^T$ on posera $N_2(f) = N_2(|f|)$. On a donc : $N_1(f) \leq N_2(f) = |m|^*(|f|)$ si f $\in \overline{\mathbb{R}}^T$. On en déduit la remarque suivante.

Remarque 6.2. (1) Si $f \in \overline{\mathbb{R}}^T$ alors on a $N_1(f) \leq N_2(f) = |m|^*(|f|)$. On en déduit la proposition 6.6.

(2) Si $A \subset T$ est m-négligeable alors A est M_m-négligeable.

(3) Si une fonction f est définie m-pp sur T alors f l'est M_m-pp.

Nous rappelons les théorèmes 5.1 et 5.2 de [12] pages 144 à 145.

Théorème 6.3. *Convergence monotone : Si $(f_n)_{n\in\mathbb{N}^*}$ est une suite croissante, (resp. décroissante) de $\ell^1(m)$ telle que $|f_n| \leq g$ pour tout $n \in \mathbb{N}^*$, et si $I^*(g) \in cc(E)$, alors $f = \sup_{n\in\mathbb{N}^*} f_n$, (resp. $\inf_{n\in\mathbb{N}^*} f_n$) $\in \ell^1(m)$, et on a $\int fm = \lim_{n\to+\infty} \int f_n m$ dans E.*

Théorème 6.4. *Convergence dominée : Si $(f_n)_{n\in\mathbb{N}^*}$ est une suite de $\ell^1(m)$ qui converge simplement vers f, s'il existe g tel que $I^*(g) \in cc(E)$ avec $|f_n| \leq g$, pour tout $n \in \mathbb{N}^*$, alors $f \in \ell^1(m)$, $\int fm = \lim_{n\to+\infty} \int f_n m$ dans E.*

Théorème 6.5. *(1) Convergence monotone : Si $(f_n)_{n\in\mathbb{N}^*}$ est une suite croissante (resp. décroissante) m-pp de $\mathsf{L}^1(m)$ telle que $|f_n| \leq g$ m-pp pour tout $n \in \mathbb{N}^*$, et si $I^*(g) \in cc(E)$, alors $f = \sup_{n\in\mathbb{N}^*} f_n$, (resp. $\inf_{n\in\mathbb{N}^*} f_n$) $\in \mathsf{L}^1(m)$, et on a $\int fm = \lim_{n\to+\infty} \int f_n m$ dans E.*

(2) Convergence dominée : Si $(f_n)_{n\in\mathbb{N}^*}$ est une suite de $\mathsf{L}^1(m)$ qui converge m-pp vers f, s'il existe g tel que $I^*(g) \in cc(E)$ avec $|f_n| \leq g$ m-pp pour tout $n \in \mathbb{N}^*$, alors $f \in \mathsf{L}^1(m)$, $\int fm = \lim_{n\to+\infty} \int f_n m$ dans E.

Dans les deux cas on a $\lim_{n\to+\infty} \int |f_n - f| m = 0$ dans E.

Preuve. La preuve est similaire à celle du théorème 3.11 : en prenant 0 comme valeurs des fonctions, dans l'union dénombrable des négligeables de l'énoncé, on se ramène aux théorèmes 6.3 et 6.4 de [12]. □

Soit $\ell^1(N_2) = \overline{H}$ dans $\{ f \in \mathbb{R}^T, N_2(f) < +\infty \}$ muni de la topologie définie par la semi-norme N_2 et soit $L^1(N_2)$={ f, définie m-pp sur T, à valeurs dans $\overline{\mathbb{R}}$, $\exists g \in \ell^1(N_2)$, f=g m-pp}. D'après la remarque 6.2, on a :

Proposition 6.6. *(1) On a $\mathsf{L}^1(|m|) = \mathsf{L}^1(N_2) \subset \mathsf{L}^1(N_1) = \mathsf{L}^1(m)$,*

(2) et $\ell^1(|m|) = \ell^1(N_2) \subset \ell^1(N_1) = \ell^1(m)$.

Remarque 6.7. L'espace $\ell^1(N_2)$ ou $\mathsf{L}^1(N_2)$ correspond à celui obtenu dans [3]. L'espace $\ell^1(N_2)$ est aussi celui obtenu, en intégrant par rapport à la semi-norme de Riesz N_2 comme dans [9] chapitre VI. L'espace $\ell^1(N_1)$ est celui obtenu dans [12] page 142. Cependant dans [12] il n'y a pas l'hypothèse E normé. L'espace $\mathsf{L}^1(N_1)$ est l'ensemble des fonctions M_m -intégrables, définies M_m-pp sur T à valeurs dans $\overline{\mathbb{R}}$.

References

[1] C. Castaing, M. Valadier, Convex Analysis and Measurable Multifunctions, *Lecture Notes in Mathematics*, Springer-Verlag, **580** (1977), 37– 50.

[2] A. Costé, Contribution à la théorie de l'intégration multivoque. *Thèse d'état*, Paris 6, (1977).

[3] N. Dinculeanu, Vector Measures, *Pergamon Press Veb-Berlin*, Vol. 95 (1967), §. 8, p. 119.

[4] C. Godet-Thobie, Multimesures et multimesures de transition. *Thèse d'état*, Montpellier, (1975).

[5] G. B. Ndiaye, Intégration par rapport à une multimesure s-compacte monotone, *Journal des sciences* - Dakar, Vol. 3 (2003), no. 1, 44–50.

[6] G. B. Ndiaye, Prolongeabilité et richesse d'une multimesure s-compacte à valeurs convexes fermées bornées, conditions de compacitée, *Journal des sciences* - Dakar, Vol. 3 (2003), no. 2, 51–55.

[7] G. B. Ndiaye, Multimesures et multimesures de Radon séquentiellement compactes. *Thèse d'état*, U. C. A. D., Dakar, (2004).

[8] R. Pallu De La Barrière, Une alternative au théorème de Banach-Dieudonné, *Seminaire d'analyse convexe*, Montpellier, Vol. II, fascicule I (1981), Exposé no. 1, I.1–I.19.

[9] R. Pallu De La Barrière, Intégration : Un nouvel itinéraire d'initiation à l'analyse mathématique, *Ellipses* (1997), Paris.

[10] R. Pallu De La Barrière, Convex Analysis. Vector and Set-valued Measures, *Publications de l'Université P. et M. Curie* **33**, I, II, III, (1977).

[11] K. Siggini, Sur les proriétées de régularité des mesures vectorielles et multivoques sur des espaces topologiques généraux. *Thèse de doctorat*, Paris 6, (1992).

[12] D. S. Thiam, Intégration dans les espaces ordonnés et intégration multivoque. *Thèse d'état*, Paris 6, (1976).

In: African Diaspora Mathematics Research Progress
Editor: Toka Diagana, pp. 31–53
ISBN 978-1-60456-204-0
© 2008 Nova Science Publishers, Inc.

Chapter 3

PSEUDO-DIFFERENTIAL OPERATORS AND COMMUTATORS IN MULTIPLIER SPACES

Mansouria Saïdani*
Department of Mathematics, Mostaganem University,
Mostaganem, B.P. 227, Algeria
Amina Lahmar-Benbernou†
Department of Mathematics, Mostaganem University,
Mostaganem, B.P. 227, Algeria
Sadek Gala‡
Department of Mathematics, Mostaganem University,
Mostaganem, B.P. 227, Algeria

Abstract

In this paper we establish the boundedness of pseudo-differential operators with symbols in the class $S^{-\frac{n\theta}{2}}_{1-a,\delta}$, $(0<\theta<1, \delta<1-\theta)$ and their commutators with BMO functions in multiplier spaces. As a consequence of this result, we extend some results on L^p by extrapolation.

AMS Subject Classification: 35S05; 42E35; 37G05; 42B30.

Keywords: Pseudo-differential operators, Sharp function, Multiplier spaces.

1 Introduction

The purpose of this paper is to establish the boundedness on the multiplier space for a class of pseudo-differential operators with symbols $\sigma(x,\xi)$ in $S^{-\frac{n\theta}{2}}_{1-\theta,\delta}$, $0<\theta<1$ and $\delta<1-\theta$ and their commutators with BMO functions. Using a Lemma 1.5 established earlier in [7] and [8], we reduce the question to establish boundedness on weighted Lebesgue spaces

*E-mail: saidaniman@yahoo.fr
†E-mail: abenbernou@yahoo.fr
‡E-mail address: sadek.gala@gmail.com

$L^2(\mathbb{R}^n, wdx)$ for weights w in the Muckenhoupt class A_2. The strategy of proof is then to adapt the results of Miller [11] on boundedness for classical pseudo-differential operators on weighted Lebesgue spaces and the technics of Fefferman and Stein [6] for the study of pseudo-differential operators on Lebesgue spaces $L^p(\mathbb{R}^n)$.

Our result is motivated by the L^p−boundedness theorem of C. Fefferman [5] who proved boundedness of pseudo-differential operators with symbols in the class $S^{-\alpha}_{1-\theta,\delta}$, with $0 < \theta < 1$, $\alpha < \frac{\theta n}{2}$ and $\delta < 1-\theta$ in L^p spaces (see also Stein [12], p. 322) or ([13], Lemma 0.5.D, p. 17). A special attention is paid also to the commutators $[b, T_\sigma]$ of a pseudo-differential operator T_σ and a functions b having bounded mean oscillation.

A pseudo-differential operator T_σ with symbol $\sigma(x,\xi)$, defined initially on the Schwartz class of testing functions $\mathcal{S}(\mathbb{R}^n)$, is given by

$$f \to T_\sigma f(x) = \int_{\mathbb{R}^n} \sigma(x,\xi)\, \widehat{f}(\xi) e^{2\pi i x\xi} d\xi, \tag{1.1}$$

with

$$\widehat{f}(\xi) = \int_{\mathbb{R}^n} f(x) e^{-2\pi i x\xi} dx$$

being the Fourier transform of f.

We shall consider in this paper pseudo-differential operators with symbols $\sigma(x,\xi)$ in the class $S^{-\frac{n\theta}{2}}_{1-\theta,\delta}$, $0 < \theta < 1$ and $\delta < 1-\theta$. A function σ belongs to $S^m_{\rho,\delta}$ if $\sigma(x,\xi)$ is a C^∞ function of $(x,\xi) \in \mathbb{R}^n \times \mathbb{R}^n$ and satisfies the differential inequalities

$$\left|\partial_x^\alpha \partial_\xi^\beta \sigma(x,\xi)\right| \le A_{\alpha,\beta} (1+|\xi|)^{m-\rho|\beta|+\delta|\alpha|},$$

for some $m \in \mathbb{R}$, $\rho, \delta \in [0,1]$ and for every n-tuplets α, β. Roughly speaking, the order $m = \frac{n}{2}(1-\rho)$ is the threshold. When $m > -\frac{n}{2}(1-\rho)$, the classical multiplier of Hardy, Hirschman and Wainger shows that pseudo-differential operators in the class $S^m_{\rho,\delta}$ fail to be continuous on L^p for some or all values of $p \neq 2$. These operators were named strongly singular by C. Fefferman [5].

In [11], Miller showed that for a symbol $\sigma(x,\xi)$ of order 0 and $1 < \gamma < \infty$, there is a constant C so that

$$(T_\sigma f)^{\neq}(x) \le C\, M_\gamma f(x),$$

where $f^{\neq}$ is the sharp function of f (see Definition in the text section).

He used this to study pseudo-differential operators of order 0 on weighted L^p spaces and ones with symbols of order m on weighted Sobolev spaces. Since then, many authors have used this technique to study pseudo-differential operators for symbols in various classes, $S^m_{\rho,\delta}$. Note that, when σ is independent of x, T_σ is a multiplier operator. One sees that this technique yields best possible estimates for pseudo-differential operators in some cases.

Before stating our result, we need to make precise the definition of the pseudo-differential operator (1.1) and the class of symbols that is used.

Definition 1.1. Let $1 < p < \infty$. A measurable function f is said to belong to the weighted L^p, $L^p(\mathbb{R}^n, wdx)$, with weight function w, if

$$\int_{\mathbb{R}^n} |f(x)|^p w(x) dx < \infty.$$

We denote the weighted L^p norm by

$$\|f\|_{L^p(w)} = \left(\int_{\mathbb{R}^n} |f(x)|^p w(x)dx \right)^{\frac{1}{p}}. \tag{1.2}$$

For $1 < p < \infty$, a positive weight function w is said to be in the class A_p if w is locally integrable and satisfies the condition

$$\sup_Q \left(\frac{1}{|Q|} \int_Q w(x)dx \right) \left(\frac{1}{|Q|} \int_Q w^{-\frac{1}{(p-1)}}(x)dx \right)^{p-1} < \infty, \tag{1.3}$$

where the supremum is taken over all cubes Q in $\mathbb{R}^n$.

Recall that a nonnegative weight $w \in L^1_{loc}(\mathbb{R}^n)$ is said to be in the Muckenhoupt class $A_1(\mathbb{R}^n)$ if

$$Mw(x) \leq \text{const } w(x) \quad \text{a.e.},$$

where M is the Hardy-Littlewood maximal function defined below. The least constant on the right-hand side of the preceding inequality is called the A_1-bound of w.

Below we list some simple, but useful properties of A_p weights.

Proposition 1.2. *1. If $w \in A_p$, $1 \leq p < \infty$, then since $w(x)^{-\frac{1}{p-1}}$ is locally integrable, when $p > 1$, and $\frac{1}{w}$ is locally bounded, when $p = 1$, we have $L^p(w(x)dx) \subset L^1_{loc}(\mathbb{R}^n)$.*

2. *Note that if w is a weight, then, by writing $1 = w^{\frac{1}{p}} w^{-\frac{1}{p}}$, Hölder's inequality implies that, for every ball B*

$$1 \leq \left(\frac{1}{|B|} \int_B w(x)dx \right) \left(\frac{1}{|B|} \int_B w(x)^{\frac{-1}{p-1}} dx \right)^{p-1}$$

when $p > 1$ and similarly for the expression that gives the A_1 condition.It follows that if $w \in A_p$, then the constant of w is ≥ 1.

3. *If $w \in A_p$, where $1 < p < \infty$, then $w^{-\frac{1}{p-1}} \in A_{p'}$, and conversely.*

4. *It is not so difficult to see that a weight $w \in A_1$ if and only if $Mw(x) \leq A_1 w(x)$ a.e.*

5. *It follows that if $w \in A_1$, then there is a constant C such that*

$$w(x) \geq \frac{C}{(1+|x|)^n}$$

for a.e. $x \in \mathbb{R}^n$. In fact, if $x \in \mathbb{R}^n$ and $R = 2\max(1, |x|)$, then

$$\frac{1}{R^n} \int_{B(R,x)} w(y)dy \geq \frac{2^{-n}}{(1+|x|)^n} \int_{B(1,0)} w(y)dy$$

so $Mw(x) \geq C(1+|x|)^{-n}$ a.e.

6. *If w is a weight and there exist two positive constants C and D such that $C \leq w(x) \leq D$, for a.e. $x \in \mathbb{R}^n$, then obviously $w \in A_p$ for $1 \leq p < \infty$.*

7. *Let $1 < p < \infty$ and $w \in A_p$. Then we have*

$$\int_{\mathbb{R}^d} \frac{w(x)}{(1+|x|)^{np}} dx < \infty .$$

The proofs of these facts are in ([9], Chapter IV) or ([12], Chapter V).

Lemma 1.3. *If $w \in A_p$, then $C_0^\infty(\mathbb{R}^n)$ is dense in $L^p(\mathbb{R}^n, wdx)$ for $1 < p < \infty$.*

Now, we recall the definition and some properties of the space we are going to use. Since then, these spaces play an important role in studying the regularity of solutions to partial differential equations; see [10], [7].

Definition 1.4. For $0 \leq r < \frac{n}{2}$, the space $\mathcal{M}\left(\dot{H}^r \to L^2\right)$ is defined as the space of $f(x) \in L^2_{loc}(\mathbb{R}^n)$ such that

$$\|f\|_{\mathcal{M}\left(\dot{H}^r \to L^2\right)} = \sup_{\|g\|_{\dot{H}^r} \leq 1} \|fg\|_{L^2} < \infty.$$

We will need the following statement established earlier in [7], for the inhomogeneous Sobolev spaces $H^r(\mathbb{R}^d)$, which shows that many operators of classical analysis are bounded in the space of functions $f \in \mathcal{M}\left(H^r \to L^2\right)$.

Lemma 1.5. *Let $f \in \mathcal{M}\left(H^r \to L^2\right)$ where $0 \leq r < \frac{n}{2}$. Suppose that T is a bounded operator on the weighted space $L^2(w)$ for every $w \in A_1(\mathbb{R}^d)$. Then $Tf \in \mathcal{M}\left(H^r \to L^2\right)$ and*

$$\|Tf\|_{\mathcal{M}(H^r \to L^2)} \leq C_1 \|f\|_{\mathcal{M}(H^r \to L^2)},$$

where the constant C_1 does not depend on f.

We have the the following result for pseudo-differential operators with symbols in $S^{-\frac{n\theta}{2}}_{1-\theta,\delta}$ acting on the space $\mathcal{M}\left(H^r \to L^2\right)$.

Theorem 1.6. *Assume $0 < \theta < 1$, $\delta < 1-\theta$ and let T_σ be the pseudo-differential operator*

$$T_\sigma f(x) = \int_{\mathbb{R}^n} \sigma(x,\xi) \widehat{f}(\xi) e^{2\pi i x\xi} d\xi.$$

with symbol $\sigma(x,\xi) \in S^{-\frac{n\theta}{2}}_{1-\theta,\delta}$. Then the operator T_σ can be extended to a bounded linear operator on $\mathcal{M}\left(H^r \to L^2\right)$.

Remark 1.7. Result similar to this in Theorem 1.6 holds for homogeneous spaces. More explicitly, if $f \in \mathcal{M}\left(\dot{H}^r \to L^2\right)$, then

$$\|T_\sigma f\|_{\mathcal{M}\left(\dot{H}^r \to L^2\right)} \leq C \|f\|_{\mathcal{M}\left(\dot{H}^r \to L^2\right)}.$$

Remark 1.8. For $0<\theta<1$, the condition $m\geq n\theta\left|\frac{1}{2}-\frac{1}{p}\right|$ guarantees the boundedness of operators with symbols in $S^{-m}_{1-\theta,\delta}$ on L^p, see Fefferman [5] (see also Stein [12], p. 322-323).

To prove this theorem, we first introduce some notations. Let Q denote any cube in $\mathbb{R}^n$ and write $|Q|$ for the Lebesgue measure of Q. For a locally integrable function f, let f_Q denote the mean value of f over Q, that is

$$f_Q=\frac{1}{|Q|}\int_Q f(x)dx.$$

We list several operators we use later :

(a) The Hardy-Littlewood maximal function, Mf, for a locally integrable function f on $\mathbb{R}^n$ by

$$Mf(x)=\sup_Q\frac{1}{|Q|}\int_Q|f(y)|dy,$$

where the supremum ranges over all cubes Q containing x.

(b) Modified maximal function of f :

$$M_\gamma f(x)=\sup_{x\in Q}\left(\frac{1}{|Q|}\int_Q|f(y)|^\gamma dy\right)^{\frac{1}{\gamma}},\quad 1\leq\gamma<\infty,$$

where the supremum is taken over all cubes Q containing x.

(c) Dyadic maximal function of f :

$$f^*(x)=\sup_{x\in Q}\frac{1}{|Q|}\int_Q|f(y)|dy$$

where the supremum is taken over all dyadic cubes Q , with sides parallel to the axes containing x.

(d) We define the sharp function of f by

$$f^{\neq}(x)=\sup_{x\in Q}\frac{1}{|Q|}\int_Q|f(y)-f_Q|dy,$$

where the supremum is taken over all cubes Q containing x. The sharp function was introduced by C. Fefferman and E.M. Stein in [6].

(e)

$$I_\alpha f=|x|^{\alpha-n}*f\qquad(0<\alpha<n).$$

It is known that I_α is a bounded operator from $L^p(\mathbb{R}^n)$ to $L^q(\mathbb{R}^n)$ with

$$\frac{1}{q}=\frac{1}{p}-\frac{\alpha}{n},\quad 1<p<\frac{n}{\alpha}.$$

Lemma 1.9. *There is a sequence* $\{\varphi_j\}_{j=0}^{\infty}$ *of functions in* $C_0^{\infty}(\mathbb{R}^n)$ *such that*

(i) $0 \leq \varphi_j(\xi) \leq 1, \quad \xi \in \mathbb{R}^n, \quad j = 0, 1, 2, ...,$

(ii) $\sum_{j=0}^{\infty} \varphi_j(\xi) = 1, \quad \xi \in \mathbb{R}^n,$

(iii) *Supp* $(\varphi_0) \subset \{\xi \in \mathbb{R}^n : |\xi| \leq 2\}$,

(iv) *Supp* $(\varphi_j) \subset \{\xi \in \mathbb{R}^n : 2^{j-1} \leq |\xi| \leq 2^{j+1}\}, \quad j = 1, 2, ...,$

(v) *for each multi-index* γ, *there is a constant* $A_\gamma > 0$ *such that*

$$\sup_{\xi \in \mathbb{R}^n} |(\partial^\gamma \varphi_j)(\xi)| \leq A_\gamma 2^{-j|\gamma|}, \qquad j = 0, 1, 2, ...$$

Lemma 1.10 ([11], Lemma 2.7). *Let* $w \in A_2(\mathbb{R}^n)$. *There is a constant* $C > 0$ *such that*

$$\|f^*\|_{L^2_w} \leq C \left\| f^{\neq} \right\|_{L^2_w} \text{ for all } f \in L^2(\mathbb{R}^n, wdx) \cap L^1(\mathbb{R}^n).$$

The following lemma is an easy consequence of result in [2].

Lemma 1.11 ([2], Theorem 1.4). *Let* $w \in A_p(\mathbb{R}^n)$. *Then there exists an* γ_0, $1 < \gamma_0 \leq p$ *such that for all* γ, $1 \leq \gamma < \gamma_0$

$$\|M_\gamma f\|_{L^p_w} \leq C_p \|f\|_{L^p_w}, \qquad 1 < p < \infty.$$

We want to obtain L^2 estimate for pseudo-differential operators. The following is a simple basic estimate but important characterization of such symbols.

Lemma 1.12 ([13], Theorem 0.5.C). *If* $p(x,\xi) \in S^0_{\rho,\delta}$ *and* $0 \leq \delta < \rho \leq 1$, *then the pseudo-differential operator with symbol* $p(x,\xi)$ *is bounded on* $L^2(\mathbb{R}^n)$.

We now observe that

$$M_p f(x) \leq M_\gamma f(x) \quad \text{for} \quad 0 < p \leq \gamma.$$

In fact for all cubes $Q \subset \mathbb{R}^n$, by Hölder' s inequality

$$\left(\frac{1}{|Q|} \int_Q |f(y)|^p \, dy \right)^{\frac{1}{p}} \leq \left(\frac{1}{|Q|} \int_Q |f(y)|^\gamma \, dy \right)^{\frac{1}{\gamma}} \leq M_\gamma f(x).$$

Given any pseudo-differential operator, we can study its kernel $K(x,y)$ defined by

$$T_\sigma f(x) = \int_{\mathbb{R}^n} K(x,y) \, f(y) dy.$$

Formally, the kernel is the following oscillatory integral

$$K(x,y) = \int_{\mathbb{R}^n} e^{2\pi i (x-y).\xi} \sigma(x,\xi) \, d\xi.$$

2 Technical Lemmas

Lemma 2.1 (Kernel estimates). *Let $\sigma(x,\xi) \in S^{-\frac{n\theta}{2}}_{1-\theta,\delta}$, and let θ be such that $0 < \theta < 1$, $0 \leq \delta < 1-\theta$. There is a constant C so that if $|x-x_0| \leq d < 1$ and $k \geq 1$, then for any $x_0 \in \mathbb{R}^n$*

$$\int_{(2^k d)^{1-\theta} \leq |y-x_0| \leq (2^{k+1}d)^{1-\theta}} |K(x,x-y) - K(x_0,x_0-y)|^2 \, dy \leq C \frac{|x-x_0|^{(1-\theta)(2m-n)}}{(2^k d)^{2m(1-\theta)}},$$

provided $m \in \mathbb{N}$ and $\frac{n}{2} < m < \frac{n}{2} + \frac{1}{1-\theta}$.

Proof. For $j = 0, 1, 2, \ldots$, we write

$$\sigma_j(x,\xi) = \sigma(x,\xi)\, \varphi_j(\xi)$$

for all $x, \xi \in \mathbb{R}^n$ and

$$K_j(x,z) = \int_{\mathbb{R}^n} e^{iz\xi} \sigma_j(x,\xi)\, d\xi$$

for all $x, z \in \mathbb{R}^n$, where $\{\varphi_j\}$ is the partition of unity constructed in Lemma 1.9. Then

$$K(x,z) = \sum_{j \in \mathbb{N}} K_j(x,z).$$

Note that

$$\left(\int_{(2^k d)^{1-\theta} \leq |y-x_0| \leq (2^{k+1}d)^{1-\theta}} |K(x,x-y) - K(x_0,x_0-y)|^2 \, dy \right)^{\frac{1}{2}}$$

$$\leq \sum_{j=0}^{\infty} \left(\int_{(2^k d)^{1-\theta} \leq |y-x_0| \leq (2^{k+1}d)^{1-\theta}} \left|K_j(x,x-y) - K_j(x_0,x_0-y)\right|^2 dy \right)^{\frac{1}{2}}. \qquad (2.1)$$

If j_0 now denotes the largest integer j so that $2^{j_0}|x-x_0| \sim 1$, it follows that

$$\sum_{j=0}^{\infty} \left(\int_{(2^k d)^{1-\theta} \leq |y-x_0| \leq (2^{k+1}d)^{1-\theta}} \left|K_j(x,x-y) - K_j(x_0,x_0-y)\right|^2 dy \right)^{\frac{1}{2}}$$

$$\leq \Sigma_1 + \Sigma_2 + \Sigma_3$$

where

$$\Sigma_1 = \sum_{j=0}^{j_0-1} \left(\int_{\left(2^k d\right)^{1-\theta} \le |y-x_0| \le \left(2^{k+1} d\right)^{1-\theta}} \left|K_j(x,x-y) - K_j(x_0,x_0-y)\right|^2 dy \right)^{\frac{1}{2}} \tag{2.2}$$

$$\Sigma_2 = \sum_{j=j_0}^{\infty} \left(\int_{\left(2^k d\right)^{1-\theta} \le |y-x_0| \le \left(2^{k+1} d\right)^{1-\theta}} \left|K_j(x,x-y)\right|^2 dy \right)^{\frac{1}{2}} \tag{2.3}$$

$$\Sigma_3 = \sum_{j=j_0}^{\infty} \left(\int_{\left(2^k d\right)^{1-\theta} \le |y-x_0| \le \left(2^{k+1} d\right)^{1-\theta}} \left|K_j(x_0,x_0-y)\right|^2 dy \right)^{\frac{1}{2}}. \tag{2.4}$$

We claim that for any $\frac{n}{2} < m < \frac{n}{2} + \frac{1}{1-\theta}$, there is a constant $C = C_{n,m}$ such that

$$\Sigma_j \le C \left(2^j d\right)^{-m(1-\theta)} |x-x_0|^{\left(m-\frac{n}{2}\right)(1-\theta)}, \quad \text{for all} \quad j = 1,2,3.$$

To prove (2.3), we multiply and divide the integrand in (2.3) by the expression $|y-x_0|^{2m}$, we control the integral in (2.3) by the product

$$\left(\int_{(2^j d)^{1-\theta} \le |y-x_0| \le (2^{j+1} d)^{1-\theta}} \left|K_j(x,x-y)\right|^2 |y-x_0|^{2m} dy \right)^{\frac{1}{2}} \tag{2.5}$$

$$\times \left(\sup_{(2^j d)^{1-\theta} \le |y-x_0| \le (2^{j+1} d)^{1-\theta}} |y-x_0|^{-2m} \right)^{\frac{1}{2}}.$$

We now note that the second expression in (2.5) is equal to a constant multiple of $\left(2^k d\right)^{-m(1-\theta)}$. To estimate the first integral in (2.5) we use the fact that

$$|x-x_0| \le d, \qquad |y-x_0| \sim \left(2^k d\right)^{1-\theta}$$

and

$$\begin{aligned} |y-x_0| &\le |y-x| + |x-x_0| \\ &\le c\,|y-x|. \end{aligned}$$

We now have that the expression in (2.3) is controlled by

$$C \sum_{j=j_0}^{\infty} \left(2^k d\right)^{-m(1-\theta)} \left(\int_{\left(2^k d\right)^{1-\theta} \le |y-x_0| \le \left(2^{k+1} d\right)^{1-\theta}} \left|K_j(x,x-y)\right|^2 |y-x|^{2m} dy \right)^{\frac{1}{2}}. \tag{2.6}$$

To estimate the above expression, we make use of the following inequality

$$|z|^{2m} \le C_m \sum_{|\gamma| \le m} |z|^{2\gamma}, \quad z \in \mathbb{R}^n.$$

We now have that the expression in (2.6) is controlled by

$$C_{n,m} \sum_{j=j_0}^{\infty} \left(2^k d\right)^{-m(1-\theta)} \sum_{|\gamma| \le m} \left(\int_{\mathbb{R}^n} |K_j(x, x-y)|^2 \, |x-y|^{2\gamma} dy \right)^{\frac{1}{2}}$$

which is equal to

$$C \sum_{j=j_0}^{\infty} \left(2^k d\right)^{-m(1-\theta)} \sum_{|\gamma| \le m} \left(\int_{\mathbb{R}^n} |(\partial^\gamma \sigma_j)(x, \xi)|^2 \, d\xi \right)^{\frac{1}{2}},$$

using Plancherel' s theorem. Now for any $|\gamma| \le m$, we use Leibnitz' s rule of differentiation to obtain

$$\begin{aligned}
\int_{\mathbb{R}^n} |(\partial^\gamma \sigma_j)(x,\xi)|^2 \, d\xi &\le \sum_{|\lambda| \le |\gamma|} C_\lambda \int_{\mathbb{R}^n} \left| 2^{-j|\gamma-\lambda|} \left(\partial_\xi^{\gamma-\lambda} \varphi\right)(2^{-j}\xi) \left(\partial_\xi^\lambda \sigma\right)(x,\xi) \right|^2 d\xi \\
&\le \sum_{|\lambda| \le |\gamma|} C_\lambda 2^{-2j|\gamma|} 2^{2j|\lambda|} \int_{2^{(j-1)} \le |\xi| \le 2^{(j+1)}} \left| \left(\partial_\xi^\lambda \sigma\right)(x,\xi) \right|^2 d\xi \\
&\le \sum_{|\lambda| \le |\gamma|} C_\lambda 2^{-2j|\gamma|} 2^{2j|\lambda|} 2^{jn} 2^{2j\left(-\frac{n\theta}{2} - (1-\theta)|\lambda|\right)} \\
&= C_n \, 2^{jn} 2^{2j\left(-\frac{n\theta}{2} - (1-\theta)|\gamma|\right)},
\end{aligned}$$

which implies

$$\begin{aligned}
& C \sum_{j=j_0}^{\infty} \left(2^k d\right)^{-m(1-\theta)} \sum_{|\gamma| \le m} \left(\int_{\mathbb{R}^n} |(\partial^\gamma \sigma_j)(x,\xi)|^2 \, d\xi \right)^{\frac{1}{2}} \\
\le \; & C \sum_{j=j_0}^{\infty} \left(2^k d\right)^{-m(1-\theta)} 2^{j\left(-\frac{n\theta}{2} - m(1-\theta) + \frac{n}{2}\right)}.
\end{aligned}$$

The choice of m assures that

$$-\frac{n\theta}{2} - m(1-\theta) + \frac{n}{2} = \left(\frac{n}{2} - m\right)(1-\theta) < 0.$$

Thus the sum above is majorized by

$$C \left(2^k d\right)^{-m(1-\theta)} 2^{j_0\left(-\frac{na}{2} - m(1-\theta) + \frac{n}{2}\right)} \le C \left(2^k d\right)^{-m(1-\theta)} |x - x_0|^{\left(m - \frac{n}{2}\right)(1-\theta)}$$

which is all we need to obtain J and L. To obtain (2.2), we shall write Σ_1 as

$$\Sigma_1 \le \Sigma_{1,1} + \Sigma_{1,2}$$

where

$$\Sigma_{1,1} = \sum_{j=0}^{j_0-1} \left(\int_{(2^k d)^{1-\theta} \le |y-x_0| \le (2^{k+1} d)^{1-\theta}} \left| K_j(x,x-y) - K_j(x_0,x-y) \right|^2 dy \right)^{\frac{1}{2}}$$

and

$$\Sigma_{1,2} = \sum_{j=0}^{j_0-1} \left(\int_{(2^k d)^{1-\theta} \le |y-x_0| \le (2^{k+1} d)^{1-\theta}} \left| K_j(x_0,x-y) - K_j(x_0,x_0-y) \right|^2 dy \right)^{\frac{1}{2}} .$$

For the first sum $\Sigma_{1,1}$, we use the fact that

$$\begin{aligned} |y-x_0| &\le |y-x| + |x-x_0| \\ &\le c\,|y-x|, \end{aligned}$$

then

$$\begin{aligned} \Sigma_{1,1} &\le \sum_{j=0}^{j_0-1} \left(2^k d\right)^{-m(1-\theta)} \left(\int_{(2^k d)^{1-\theta} \le |y-x_0| \le (2^{k+1} d)^{1-\theta}} \frac{\left| K_j(x,x-y) - K_j(x_0,x-y) \right|^2}{|y-x_0|^{-2m}} dy \right)^{\frac{1}{2}} \\ &\le c \sum_{j=0}^{j_0-1} \left(2^k d\right)^{-m(1-\theta)} \left(\int_{(2^k d)^{1-\theta} \le |y-x_0| \le (2^{k+1} d)^{1-\theta}} \frac{\left| K_j(x,x-y) - K_j(x_0,x-y) \right|^2}{|y-x|^{-2m}} dy \right)^{\frac{1}{2}} \\ &\le c \sum_{j=0}^{j_0-1} \left(2^k d\right)^{-m(1-\theta)} \sum_{|\gamma| \le m} \left(\int_{\mathbb{R}^n} \sup_{\eta} \left| \partial_\eta \partial_\xi^\gamma \sigma_j(\eta,\xi) \right|^2 d\xi \right)^{\frac{1}{2}} |x-x_0| \\ &\le C \sum_{j=0}^{j_0-1} \left(2^k d\right)^{-m(1-\theta)} 2^{j\left(-\frac{n\theta}{2} - m(1-\theta) + \frac{n}{2} + 1\right)} |x-x_0| . \end{aligned}$$

The choice of m and the fact that $2^{j_0}|x-x_0| \sim 1$, it follows that

$$\Sigma_{1,1} \le C\left(2^k d\right)^{-m(1-\theta)} |x-x_0|^{\left(m-\frac{n}{2}\right)(1-\theta)} .$$

Now we claim the second sum $\Sigma_{1,2}$ also be made $\le C\left(2^k d\right)^{-m(1-\theta)} |x-x_0|^{\left(m-\frac{n}{2}\right)(1-\theta)}$. Indeed, this second sum $\Sigma_{1,2}$ can be bounded by

$$\sum_{j=0}^{j_0-1} \left(2^k d\right)^{-m(1-\theta)} \left(\int_{(2^k d)^{1-\theta} \le |y-x_0| \le (2^{k+1} d)^{1-\theta}} \frac{\left| K_j(x_0,x-y) - K_j(x_0,x_0-y) \right|^2}{|y-x_0|^{-2m}} dy \right)^{\frac{1}{2}} .$$

The cancellation of $K_j(x_0,x-y)-K_j(x_0,x_0-y)$ must play a role. Define

$$\begin{aligned}\widetilde{K}_j(x_0,y) &= K_j(x_0,x-y)-K_j(x_0,x_0-y)\\ \widetilde{\sigma}_j(x_0,\xi) &= \mathcal{F}\left(\widetilde{K}_j(x_0,y)\right)(\xi)=\sigma_j(x_0,\xi)\left(e^{i(x-x_0,\xi)}-1\right).\end{aligned}$$

Our task consists now in obtaining an estimate for the derivatives of $\widetilde{\sigma}_j(x_0,\xi)$ which is good enough, namely

$$\left|\partial_\xi^\beta \sigma_j(x_0,\xi)\right|\leq C2^{j\left(-\frac{n\theta}{2}-|\beta|(1-\theta)+\frac{n}{2}\right)}$$

To prove this, we first observe that, for all $\xi\in$Supp $(\widetilde{\sigma}_j)$

$$\left|\partial_\xi^\gamma\left(e^{i(x-x_0,\xi)}-1\right)\right|\leq C|x-x_0|\,2^{j(1-|\gamma|)}\qquad (j<j_0)$$

(this is obvious when $\gamma=0$ because

$$|(x-x_0).\xi|\leq 2^{j+1}|x-x_0|,$$

and when $|\gamma|>0$, we have

$$\left|\partial_\xi^\gamma\left(e^{i(x-x_0,\xi)}\right)\right|=|x-x_0|^{|\gamma|}$$

By the Leibniz rule

$$\begin{aligned}\sup_{|\alpha|\leq m}\left|\partial_\xi^\alpha\widetilde{\sigma}_j(x_0,\xi)\right| &\leq \sum_{|\gamma|+|\beta|\leq m}\left|\partial_\xi^\beta\sigma_j(x_0,\xi)\partial_\xi^\gamma\left(e^{i(x-x_0,\xi)}-1\right)\right|\\ &\leq C\sum_{|\gamma|+|\beta|\leq m}|x-x_0|^{|\gamma|}2^{j\left(-\frac{n\theta}{2}-|\beta|(1-\theta)+\frac{n}{2}\right)}.\end{aligned}$$

Therefore, it follows immediately from the Leibniz formula that one has for a suitable constant c :

$$\begin{aligned}\Sigma_{1,2} &\leq c\sum_{j=0}^{j_0-1}\left(2^k d\right)^{-m(1-\theta)}\left(\int_{\mathbb{R}^n}\sum_{|\gamma|+|\beta|\leq m}\left|\partial_\xi^\beta\sigma_j(x_0,\xi)\partial_\xi^\gamma\left(e^{i(x-x_0,\xi)}-1\right)\right|^2 d\xi\right)^{\frac{1}{2}}\\ &\leq c\sum_{j=0}^{j_0-1}\left(2^k d\right)^{-m(1-\theta)}\left(\sum_{\substack{|\gamma|+|\beta|\leq m\\ |\gamma|\neq 0}}|x-x_0|^{|\gamma|}2^{j\left(-\frac{n\theta}{2}-|\beta|(1-\theta)+\frac{n}{2}\right)}\right)\\ &+c\sum_{j=0}^{j_0-1}\left(2^k d\right)^{-m(1-\theta)}\left(|x-x_0|2^{j\left(-\frac{n\theta}{2}-|\beta|(1-\theta)+\frac{n}{2}+1\right)}\right).\end{aligned}$$

But $|x-x_0|\leq d\leq\frac{1}{2}$ and $m<\frac{n}{2}+\frac{1}{1-\theta}$, it is easily seen that the second sum above dominates the first one. Since $2^{j_0}|x-x_0|\sim 1$, this second term is readily seen to be bounded by

$$c\left(2^k d\right)^{-m(1-\theta)}|x-x_0|^{\left(m-\frac{n}{2}\right)(1-\theta)},$$

as desired. This completes the proof. □

We shall recall one more fact about symbols in $S^m_{\rho,\delta}$, namely that the convolution kernels are essentially compactly supported.

Lemma 2.2. *Suppose T_σ is a pseudo-differential operator whose symbol $\sigma(x,\xi)$ belongs to $S^0_{\rho,\delta}$, $0<\rho<1$, and let*

$$K(x,x-y)=\int_{\mathbb{R}^n} e^{2\pi i(x-y).\xi}\sigma(x,\xi)\,d\xi$$

in distribution sense. Then for any integer $N\geq 1$, one can find a constant $C=C_N$ so that

$$|K(x,x-y)|\leq C|x-y|^{-2N} \tag{2.7}$$

for all $|x-y|\geq 1$.

Proof. For simplicity, we shall consider only the case of symbols $\sigma=\sigma(\xi)$, since in the general case $\sigma(x,\xi)$ we can consider x as a parameter. Setting $z=x-y$, we shall evaluate the kernel

$$K(z)=\int_{\mathbb{R}^n} e^{2\pi iz\xi}\sigma(\xi)\,d\xi.$$

Note that $K(z)$ is a function of z lies in $\mathcal{S}$. In fact, for any multi-indices α and β

$$\begin{aligned}\left|z^\alpha\left\{\partial^\beta K\right\}(z)\right| &= c_{\alpha\beta}\left|\int_{\mathbb{R}^n} e^{iz\xi}\sigma(\xi)\,\xi^\beta\left(\frac{\partial}{\partial\xi}\right)^\alpha d\xi\right|\\ &\leq c_{\alpha\beta}\int_{\mathbb{R}^n}\left|\left(\frac{\partial}{\partial\xi}\right)^\alpha\left[\sigma(\xi)\,\xi^\beta\right]\right|d\xi\\ &\qquad\text{(integration by parts)}\\ &\leq c_{\alpha\beta},\end{aligned}$$

with $c_{\alpha\beta}$ independent of z. The rapid decrease in ξ of $\sigma(\xi)$ justifies the differentiation under the integral sign and the integration by parts in the calculation above. Hence,

$$\sup_{z\in\mathbb{R}^n}\left|z^\alpha\left\{\partial^\beta K\right\}(z)\right|\leq c_{\alpha\beta}.$$

Now integration by parts gives

$$K(z)=\int_{\mathbb{R}^n} e^{iz\xi}\sigma(\xi)\,d\xi=(-1)^N|z|^{-2N}\int_{\mathbb{R}^n}\Delta_\xi^N\sigma(\xi)\,e^{iz\xi}d\xi$$

if $z\neq 0$. Since

$$\left|\Delta_\xi^N\sigma(\xi)\right|\leq C_N\left(1+|\xi|^2\right)^{-N\rho}$$

is integrable when $N>\frac{n}{2\rho}$, we conclude that (2.7) holds for $|z|\geq 1$ and this completes the proof of the Lemma 2.2. □

3 An $L^2(\mathbb{R}^n, wdx)$ Theorem

After these preliminaries, we state the first main result which constitutes the main part of the proof of theorem 1.6.

Theorem 3.1. *Suppose* $2 \leq \gamma < \infty$ *and let* T_σ *be a pseudo-differential operator with symbol* $\sigma(x,\xi) \in S^{-\frac{\theta n}{2}}_{1-\theta,\delta}$. *Then there is a constant* $C > 0$ *such that for all* $x_0 \in \mathbb{R}^n$ *and for all* $f \in \mathcal{D}(\mathbb{R}^n)$,

$$(T_\sigma f)^{\neq}(x_0) \leq CM_\gamma f(x_0). \tag{3.1}$$

Remark 3.2. Using the density of $\mathcal{D}(\mathbb{R}^n)$ in $L^2(\mathbb{R}^n, wdx)$, it is sufficient to prove Theorem 3.1 for functions f in $\mathcal{D}(\mathbb{R}^n)$.

Proof. We imitate the arguments given in Fefferman-Stein [6]. For arbitrary $x_0 \in \mathbb{R}^n$, we let Q be a cube centered at x_0 and diameter d. We distinguish two cases.

Case 1. $d = \text{diam } Q \leq 1$.

Let Q^* be the cube with the same center as Q with whose sides parallel to those of Q and with diameter $d^{1-\theta}$. Given $f \in \mathcal{D}(\mathbb{R}^n)$ and decompose f as

$$f = f_1 + f_2$$

where f_1 is defined by

$$f_1(x) = \begin{cases} f(x) & \text{in } Q^* \\ 0 & \text{elsewhere} \end{cases}$$

belongs to $\mathcal{D}(\mathbb{R}^n)$. Furthermore, by linearity

$$T_\sigma f = T_\sigma f_1 + T_\sigma f_2.$$

We have

$$\begin{aligned} & \frac{1}{|Q|}\int_Q |T_\sigma f(x) - T_\sigma f_2(x_0)|\,dx \\ \leq\ & \frac{1}{|Q|}\int_Q |T_\sigma f_1(x)|\,dx + \frac{1}{|Q|}\int_Q |T_\sigma f_2(x) - T_\sigma f_2(x_0)|\,dx \\ =\ & I + II. \end{aligned}$$

To estimate I, we note

$$\sigma(x,\xi) = |\xi|^{-\frac{n\theta}{2}} \sigma(x,\xi)\,|\xi|^{\frac{n\theta}{2}} = p(x,\xi)\,|\xi|^{-\frac{n\theta}{2}}.$$

It is easily verified that $p(x,\xi) \in S^0_{1-\theta,\delta}$ and it follows by Lemma 1.12, that the pseudo-differential operator with symbol $p(x,\xi)$ is bounded on $L^2(\mathbb{R}^n)$. We write also

$$\begin{aligned} T_\sigma f_1 &= I_{\frac{n\theta}{2}} I_{(-\frac{n\theta}{2})} T_\sigma f_1 \\ &= I_{\frac{n\theta}{2}}(g) \end{aligned}$$

where I_θ is the Riesz potential and

$$g = I_{\left(-\frac{n\theta}{2}\right)} T_\sigma f_1.$$

If we define q such that

$$\frac{1}{q} = \frac{1}{2} - \frac{\frac{n\theta}{2}}{n} = \frac{1}{2} - \frac{\theta}{2},$$

then by using the Hardy-Littlewood-Sobolev theorem of fractional integration, we get

$$\|T_\sigma f_1\|_{L^q} = \left\| I_{\frac{n\theta}{2}}(g) \right\|_{L^q} \leq C \|g\|_{L^2}.$$

Thus, by Hölder' s inequality

$$\begin{aligned}
\frac{1}{|Q|} \int_Q |T_\sigma f_1(x)| \, dx &\leq \left(\frac{1}{|Q|} \int_Q |T_\sigma f_1(x)|^q \, dx \right)^{\frac{1}{q}} \\
&\leq c \frac{1}{|Q|^{\frac{1}{q}}} \left(\int_{\mathbb{R}^n} |T_\sigma f_1(x)|^q \, dx \right)^{\frac{1}{q}} \\
&\leq C \frac{1}{|Q|^{\frac{1}{q}}} \left(\int_{\mathbb{R}^n} |g(x)|^2 \, dx \right)^{\frac{1}{2}}.
\end{aligned}$$

Since $g = I_{\left(-\frac{n\theta}{2}\right)} T_\sigma f_1$ is bounded on $L^2(\mathbb{R}^n)$, one has

$$\begin{aligned}
\frac{1}{|Q|} \int_Q |T_\sigma f_1(x)| \, dx &\leq C \frac{1}{|Q|^{\frac{1}{q}}} \left(\int_{\mathbb{R}^n} |f_1(x)|^2 \, dx \right)^{\frac{1}{2}} \\
&\leq C \frac{1}{|Q|^{\frac{1}{q}}} \left(\int_{Q^*} |f(x)|^2 \, dx \right)^{\frac{1}{2}} \\
&\leq C \frac{|Q^*|^{\frac{1}{2}}}{|Q|^{\frac{1}{q}}} M_2 f(x_0) \\
&\leq C \, M_\gamma f(x_0).
\end{aligned}$$

Now let

$$a_Q = \int_{\mathbb{R}^n} K(x_0, x_0 - y) \, f_2(y) dy.$$

Since

$$T_\sigma f_2(x) - a_Q = \int_{\mathbb{R}^n} \left[K(x, x-y) - K(x_0, x_0 - y) \right] f_2(y) dy,$$

we get for $x \in Q$

$$
\begin{aligned}
& |T_\sigma f_2(x) - a_Q| \\
= & \left| \int_{\mathbb{R}^n} [K(x, x-y) - K(x_0, x_0 - y)] f_2(y) dy \right| \\
\leq & \left| \int_{(Q^*)^c} [K(x, x-y) - K(x_0, x_0 - y)] f(y) dy \right| \\
= & \left| \sum_{k=1}^{\infty} \int_{(2^k d)^{1-\theta} \leq |y - x_0| \leq (2^{k+1} d)^{1-\theta}} [K(x, x-y) - K(x_0, x_0 - y)] f(y) dy \right| \\
\leq & \sum_{k=1}^{\infty} \left(\int_{(2^k d)^{1-\theta} \leq |y - x_0| \leq (2^{k+1} d)^{1-\theta}} |K(x, x-y) - K(x_0, x_0 - y)|^2 dy \right)^{\frac{1}{2}} \\
& \times \left(\int_{(2^k d)^{1-\theta} \leq |y - x_0| \leq (2^{k+1} d)^{1-\theta}} |f(y)|^2 dy \right)^{\frac{1}{2}}.
\end{aligned}
$$

By using Lemma 2.1 one has for $|x - x_0| \leq d$,

$$
\begin{aligned}
|T_\sigma f_2(x) - a_Q| & \leq C \sum_{k=1}^{\infty} \left[d^{(1-\theta)(m - \frac{n}{2})} \left(2^k d\right)^{-m(1-\theta)} \right] \left[\left(2^k d\right)^{\frac{n(1-\theta)}{2}} M_2 f(x_0) \right] \\
& \leq C\, M_2 f(x_0) \sum_{k=1}^{\infty} 2^{k[(\frac{n}{2} - m)(1-\theta)]} \qquad \left(\frac{n}{2} < m\right) \\
& \leq C\, M_2 f(x_0) \qquad (2 \leq \gamma < \infty) \\
& \leq C\, M_\gamma f(x_0)
\end{aligned}
$$

for $x_0 \in Q$. We have showed that

$$
\begin{aligned}
& \frac{1}{|Q|} \int_Q |T_\sigma f(x) - a_Q| dx \\
\leq & \frac{1}{|Q|} \int_Q |T_\sigma f_1(x)| dx + \frac{1}{|Q|} \int_Q |T_\sigma f_2(x) - a_Q| dx \\
\leq & C\, M_\gamma f(x_0).
\end{aligned}
$$

Taking the supremum of the left side over all cubes Q containing x_0, we get that

$$
(T_\sigma f)^{\neq}(x_0) \leq C\, M_\gamma f(x_0)
$$

Case 2. Suppose now $d = \operatorname{diam} Q > 1$. Let $2Q$ be the cube centered at x_0 with twice the diameter of Q. Again decompose f as

$$f = f_1 + f_2$$

where f_1 is the restriction of f to $2Q$. Therefore,

$$T_\sigma f = T_\sigma f_1 + T_\sigma f_2.$$

Then for $T_\sigma f_1$, since T_σ is bounded on $L^2(\mathbb{R}^n)$, one has

$$\|T_\sigma f_1\|_{L^2(\mathbb{R}^n)} \leq C\|f_1\|_{L^2(\mathbb{R}^n)}.$$

Moreover, by Cauchy-Schwarz inequality,

$$\begin{aligned}
\int_Q |T_\sigma f_1(x)|\,dx &\leq \|T_\sigma f_1\|_{L^2(\mathbb{R}^n)} |Q|^{\frac{1}{2}} \\
&\leq c\ |Q|^{\frac{1}{2}} \left(\int_{2Q} |f(x)|^2\,dx \right)^{\frac{1}{2}} \\
&\leq C|Q| M_2 f(x_0) \quad (2 \leq \gamma < \infty) \\
&\leq C|Q| M_\gamma f(x_0).
\end{aligned}$$

It follows from this inequality that

$$\frac{1}{|Q|} \int_Q |T_\sigma f_1(x)|\,dx \leq C M_\gamma f(x_0)$$

for all $2 \leq \gamma < \infty$ and for $x_0 \in 2Q$.

About $T_\sigma f_2(x)$, we use the rapid decrease of $K(x,y)$. Indeed, by Lemma 2.2, there is a positive constant C_n such that for all $x \in 2Q$,

$$\begin{aligned}
|T_\sigma f_2(x)| &= \left| \int_{\mathbb{R}^n} K(x, x-y)\, f_2(y) dy \right| \\
&= \left| \int_{\mathbb{R}^n \setminus 2Q} K(x, x-y)\, f_2(y) dy \right| \\
&\leq C_{2N} \int_{|y-x_0|>2d} \frac{|f(y)|}{|x-y|^{2N}} dy.
\end{aligned} \tag{3.2}$$

However when $|x - x_0| \leq d$,

$$|x-y| \geq |y-x_0| - |x-x_0| > 2d - d = d,$$

using then the fact that $n \geq 3$ together with (3.2), we get

$$\begin{aligned}
|T_\sigma f_2(x)| &\leq C \sum_{k=1}^{\infty} \int_{2^k d<|y-x_0|<2^{k+1}d} \frac{|f(y)|}{|x-y|^{2n}} dy \\
&\leq C \sum_{k=1}^{\infty} \left(2^k d\right)^{-2n} \int_{|y-x_0|\leq 2^{k+1}d} |f(y)|\, dy \\
&\leq C\, Mf(x_0) \\
&\leq C\, M_\gamma f(x_0)
\end{aligned}$$

for all $1<\gamma<\infty$ and for $x_0 \in 2Q$. Combining all these estimates, and since Q is arbitrary, the desired result (3.1) follows readily. Hence, the proof of Theorem 3.1 is complete. □

We are ready to prove a basic result about pseudo-differential operators.

Theorem 3.3. *If $w \in A_1\left(\mathbb{R}^d\right)$, then the operator T_σ with symbol $\sigma(x,\xi) \in S^{-\frac{n\theta}{2}}_{1-\theta,\delta}$, initially defined on $\mathcal{D}$, extends to a bounded operator from $L^2\left(\mathbb{R}^d, wdx\right)$ to itself.*

To prove the theorem, it suffices to show that

$$\|T_\sigma f\|_{L^2_w} \leq C\|f\|_{L^2_w}, \quad \text{whenever } f \in \mathcal{D}(\mathbb{R}^n),$$

with C independent of f.

Proof. We prove this in the same way as was used by [[11], Theorem 2.12]. If $f \in \mathcal{D}(\mathbb{R}^n)$, then since $T_\sigma f \in L^2(\mathbb{R}^n, wdx) \cap L^1(\mathbb{R}^n)$

$$\begin{aligned}
\|T_\sigma f\|_{L^2_w} &\leq \left\|(T_\sigma f)^*\right\|_{L^2_w} \leq C\left\|(T_\sigma f)^{\neq}\right\|_{L^2_w} \\
&\leq C\left\|M_\gamma f\right\|_{L^2_w} \quad \text{if } 2<\gamma<\infty \\
&\leq C\|f\|_{L^2_w} \quad \text{if } 1<\gamma<2.
\end{aligned}$$

(cf. Theorem 2.12 in [11]). Because of lemma 1.3, we can extend T_σ to a bounded operator on $L^2(\mathbb{R}^n, wdx)$. □

Theorem 1.6 follows by combining Theorem 3.3 with Lemma 1.5.

Lemma 3.4 (Rubio de Francia' s Extrapolation Theorem). *Assume T is a bounded linear operator in $L^2(\mathbb{R}^n, wdx)$ for all weights $w \in A_2(\mathbb{R}^n)$. Then T is bounded in $L^q(\mathbb{R}^n)$ for all $1<q<\infty$.*

From Theorem 3.3 and Lemma 3.4, we immediately get the following corollary.

Corollary 3.5. *Let T_σ be a pseudo-differential operator with symbol $\sigma(x,\xi) \in S^{-\frac{n\theta}{2}}_{1-\theta,\delta}$. Then there is a constant $C>0$ such that*

$$\|T_\sigma f\|_{L^q} \leq C\|f\|_{L^q}$$

for all $f \in \mathcal{D}(\mathbb{R}^n)$ and for all $1<q<\infty$.

4 Commutators with *BMO* Functions

We first recall the definition of the space *BMO* on $\mathbb{R}^n$ which was introduced by John and Nirenberg while studying PDE' s.

Definition 4.1. A measurable function b on $\mathbb{R}^n$ is in the space *BMO* if

$$\|b\|_{BMO} = \sup_{Q} \frac{1}{|Q|} \int_{Q} |b(y) - b_Q|\, dy < \infty$$

where the supremum is taken over all cubes Q in $\mathbb{R}^n$ with sides parallel to the axes, and

$$b_Q = \frac{1}{|Q|} \int_{Q} b(y) dy.$$

Here and in what follows $|E|$ denotes Lebesgue measure of a measurable set $E \subset \mathbb{R}^n$.

Note that $\|b\|_{BMO}$ is not a norm, but a seminorm, with the property that

$$\|b\|_{BMO} = 0 \quad \text{if and only if} \quad b = const.$$

It is thus natural to consider the quotient space $BMO/\mathbb{R}$ with the norm induced by $\|.\|_{BMO}$. Then $BMO/\mathbb{R}$ is a Banach space, which will also be denoted *BMO* for simplicity. We easily see that $L^\infty(\mathbb{R}^n) \subset BMO$ with continuous injection. A typical example of an unbounded function in *BMO* is a function of the form $\log|P|$ where P is an arbitrary polynomial.

We give one property of the *BMO* space in the following lemma.

Lemma 4.2. *Suppose that there exists an $A > 0$ such that for all cubes Q in $\mathbb{R}^n$ there exists a constant C_Q such that*

$$\sup_{Q} \frac{1}{|Q|} \int_{Q} |b(y) - C_Q|\, dy \leq A. \tag{4.1}$$

Then $b \in BMO(\mathbb{R}^n)$ and

$$\|b\|_{BMO} \leq 2A.$$

Proof. Note that

$$\begin{aligned} |b - C_Q| &\leq |b - b_Q| + |b_Q - C_Q| \\ &\leq |b - C_Q| + \frac{1}{|Q|} \int_{Q} |b(t) - C_Q|\, dt. \end{aligned}$$

Averaging over Q and using (4.1), we obtain that

$$\|b\|_{BMO} \leq 2A.$$

□

Lemma 4.3 (John-Nirenberg Lemma). *Let $1 \leq p < \infty$. Then $b \in BMO$ if and only if*

$$\sup_Q \frac{1}{|Q|} \int_Q |b(y) - b_Q|^p \, dy \leq C \|b\|_{BMO}^p.$$

Proof. For the proof of this lemma, see [1]. □

The next result is a sharp version of the sufficient condition in [4], Theorem 1.6.

Theorem 4.4. *Let T_σ be a pseudo-differential operator with symbol $\sigma(x,\xi) \in S_{1-\theta,\delta}^{-\frac{n\theta}{2}}$ and let $b \in BMO$, fixed. Then the linear commutator*

$$[b, T_\sigma](f) = bT_\sigma(f) - T_\sigma(bf)$$

maps continuously $\mathcal{M}\left(\dot{H}^r \to L^2\right)$ to itself $\left(0 \leq r < \frac{n}{2}\right)$. Moreover,

$$\|[b, T_\sigma](f)\|_{\mathcal{M}\left(\dot{H}^r \to L^2\right)} \leq C \|b\|_{BMO} \|f\|_{\mathcal{M}\left(\dot{H}^r \to L^2\right)}$$

for some constant $C \geq 0$ independent of b and f.

The proof of Theorem 4.4 is based on the following a priori estimate.

Proposition 4.5. *Let T_σ be a pseudo-differential operator satisfying the hypothesis of Theorem 4.4 and let $b \in BMO$, fixed and assume that T_σ is continuous on L^q for $q > 1$. Then, there exists a constant $c > 0$ independent of b and f such that*

$$[b, T_\sigma](f)^{\neq}(x_0) \leq c \|b\|_{BMO} \left(M_\gamma(f)(x_0) + M_\gamma(T_\sigma f)(x_0)\right)$$

for all $x_0 \in \mathbb{R}^n$ and every $f \in \mathcal{D}(\mathbb{R}^n)$ $(1 < \gamma < \infty)$.

Proof. Given $f \in \mathcal{D}(\mathbb{R}^n)$ and a cube Q centered at x_0 and has diameter d. Let Q^* be the cube with the same center as Q with sides parallel to those of Q, and with diameter $d^{1-\theta}$. As in the proof of Theorem 3.1 decompose f as

$$f = f_1 + f_2$$

where f_1 is the restriction of f to Q^*. Now let b_Q, the average of the function b over the cube Q. Then

$$\begin{aligned}
[b, T_\sigma](f) &= (b - b_Q)T_\sigma(f) - T_\sigma((b - b_Q)f) \\
&= (b - b_Q)T_\sigma(f) - T_\sigma((b - b_Q)f_2) - T_\sigma((b - b_Q)f_1) \\
&= J_3 + J_2 + J_1.
\end{aligned}$$

To estimate the sharp function of the commutator we employ the idea of Str ömberg (see [14]) which consists of expressing $[b, T_\sigma](f)$ as a sum of integral operators and estimating their sharp functions. Now for arbitrary $\gamma \in (1,\infty)$ and $q \in (1,\gamma)$, we obtain

$$\begin{aligned} S_1(x_0,Q) &= \frac{1}{|Q|}\int\limits_Q \left|J_1(x)-(J_1)_Q\right|dx \\ &\leq \frac{2}{|Q|}\int\limits_Q |T_\sigma(b-b_Q)f_1(x)|\,dx \\ &\leq \frac{2|Q|^{1-\frac{1}{q}}}{|Q|}\left(\int\limits_Q |T_\sigma(b-b_Q)f_1(x)|^q\,dx\right)^{\frac{1}{q}}. \end{aligned}$$

Since $q>1$, T_σ is bounded on $L^q(\mathbb{R}^n)$. Therefore, by Hölder' s inequality

$$\begin{aligned} S_1(x_0,Q) &\leq C\left(\frac{1}{|Q|}\int\limits_{\mathbb{R}^n} |(b(x)-b_Q)f_1(x)|^q\,dx\right)^{\frac{1}{q}} \\ &\leq C\left(\frac{1}{|Q|}\int\limits_{Q^*} |(b(x)-b_Q)f(x)|^q\,dx\right)^{\frac{1}{q}} \\ &\leq C\left(\frac{1}{|Q^*|}\int\limits_{Q^*} |f(x)|^\gamma dx\right)^{\frac{1}{\gamma}}\left(\frac{1}{|Q^*|}\int\limits_{Q^*} |(b(x)-b_Q)|^{\frac{\gamma q}{\gamma-q}}\,dx\right)^{\frac{\gamma-q}{\gamma q}}. \end{aligned}$$

Now we start with the following observation. For any cubes $Q^*,Q\subset\mathbb{R}^n$ such that $Q^*\subset Q$ and any $b\in BMO(\mathbb{R}^n)$, we then have

$$\begin{aligned} |b_Q-b_{Q^*}| &= \left|\frac{1}{|Q^*|}\int\limits_{Q^*}(b(x)-b_Q)dx\right| \\ &\leq \frac{1}{|Q^*|}\int\limits_{Q^*}|(b(x)-b_Q)|\,dx \\ &\leq \left(\frac{|Q|}{|Q^*|}\right)\frac{1}{|Q|}\int\limits_Q |(b(x)-b_Q)|\,dx \\ &\leq \left(\frac{|Q|}{|Q^*|}\right)\|b\|_{BMO}. \end{aligned}$$

From this observation, it follows taht for any two closed cubes $Q^*,Q\subset\mathbb{R}^n$ such that int $(Q^*\cap Q)\neq\emptyset$, there exists a constant $C(Q^*,Q)>0$such that

$$|b_Q-b_{Q^*}|\leq C(Q^*,Q)\,\|b\|_{BMO}$$

for all $b \in BMO(\mathbb{R}^n)$. Then

$$\begin{aligned}
&\int_{Q^*} |(b(x)-b_Q)|^{\frac{\gamma q}{\gamma-q}}\,dx \\
\leq\ & 2^{\frac{\gamma q}{\gamma-q}-1}\left(\int_{Q^*}|b(x)-b_{Q^*}|^{\frac{\gamma q}{\gamma-q}}\,dx+\int_{Q^*}|b_{Q^*}-b_Q|^{\frac{\gamma q}{\gamma-q}}\,dx\right) \\
\leq\ & c(\gamma,q)\left(|Q^*|\frac{1}{|Q^*|}\int_{Q^*}|(b(x)-b_{Q^*})|^{\frac{\gamma q}{\gamma-q}}\,dx+|Q^*|\,c(n)\,\|b\|_{BMO}^{\frac{\gamma q}{\gamma-q}}\right) \\
\leq\ & c(\gamma,q,n)\,|Q^*|\,\|b\|_{BMO}^{\frac{\gamma q}{\gamma-q}}.
\end{aligned}$$

Therefore, $S_1(x_0,Q)\leq c\,\|b\|_{BMO}\left(\frac{1}{|Q^*|}\int_{Q^*}|f(x)|^{\gamma}\,dx\right)^{\frac{1}{\gamma}}\leq c\,\|b\|_{BMO}M_{\gamma}(f)(x_0)$.

To estimate the sharp function of $J_2(x)$, we have

$$S_2(x_0,Q)=\frac{1}{|Q|}\int_Q\left|J_2(x)-(J_2)_Q\right|dx\leq\frac{2}{|Q|}\int_Q|J_2(x)-J_2(x_0)|\,dx$$

and the integrand satisfies for $|x-x_0|\leq d<1$

$$\begin{aligned}
|J_2(x)-J_2(x_0)| \leq\ & \int_{(Q^*)^c}|K(x,x-y)-K(x_0,x_0-y)|\,|b(y)-b_Q|\,|f(y)|\,dy \\
\leq\ & \sum_{j=0}^{\infty}\int_{(2^jd)^{1-\theta}\leq|y-x_0|\leq(2^{j+1}d)^{1-\theta}}|K(x,x-y)-K(x_0,x_0-y)|\,|b(y)-b_Q|\,|f(y)|\,dy \\
\leq\ & C\sum_{j=1}^{\infty}\left(\int_{(2^jd)^{1-\theta}\leq|y-x_0|\leq(2^{j+1}d)^{1-\theta}}|K(x,x-y)-K(x_0,x_0-y)|^2\,dy\right)^{\frac{1}{2}} \\
& \times\left(\int_{|y-x_0|\leq(2^{j+1}d)^{1-\theta}}|b(y)-b_Q|^2\,|f(y)|^2\,dy\right)^{\frac{1}{2}} \\
\leq\ & C\sum_{j=1}^{\infty}(2^jd)^{-(m-\frac{n}{2})(1-\theta)}\left(\frac{1}{\left|B\left(x_0,(2^{j+1}d)^{1-\theta}\right)\right|}\int_{|y-x_0|\leq(2^{j+1}d)^{1-\theta}}|f(y)|^{\gamma}\,dy\right)^{\frac{1}{\gamma}} \\
& \times\left(\frac{1}{\left|B\left(x_0,(2^{j+1}d)^{1-\theta}\right)\right|}\int_{|y-x_0|\leq(2^{j+1}d)^{1-\theta}}|b(y)-b_Q|^s\,dy\right)^{\frac{1}{s}} \\
\leq\ & C\,\|b\|_{BMO}M_{\gamma}(f)(x_0).
\end{aligned}$$

where $\frac{1}{\gamma}+\frac{1}{s}=\frac{1}{2}$. Hence,

$$S_2(x_0,Q) \leq c\,\|b\|_{BMO} M_\gamma(f)(x_0).$$

Finally, to estimate the sharp function of $J_3(x)$,

$$\begin{aligned}
S_3(x_0,Q) &= \frac{1}{|Q|}\int_Q \left|J_3(x)-(J_3)_Q\right| dx \\
&\leq \frac{2}{|Q|}\int_Q |b(x)-b_Q|\,|T_\sigma f(x)|\,dx \\
&\leq 2\left(\frac{1}{|Q|}\int_Q |b(x)-b_Q|^{p'}\,dx\right)^{\frac{1}{p'}}\left(\frac{1}{|Q|}\int_Q |T_\sigma f(x)|^p\,dx\right)^{\frac{1}{p}} \\
&\leq c(p)\,\|b\|_{BMO} M_\gamma(T_\sigma f)(x_0).
\end{aligned}$$

Summing up $S_1(x_0,Q)$, $S_2(x_0,Q)$ and $S_3(x_0,Q)$ and taking the supremum with respect to Q and rendering in account the arbitrary of the point x_0 we finally obtain the inequality

$$[b,T_\sigma](f)^{\neq}(x_0) \leq c\,\|b\|_{BMO}\left(M_\gamma(f)(x_0)+M_\gamma(T_\sigma f)(x_0)\right).$$

□

We are now ready to prove a basic result about commutator.

Proof. Let $f \in \mathcal{D}(\mathbb{R}^n)$. Then, since $[b,T_\sigma](f) \in L^2(\mathbb{R}^n, wdx) \cap L^1(\mathbb{R}^n)$

$$\begin{aligned}
\|[b,T_\sigma](f)\|_{L^2_w} &\leq \left\|([b,T_\sigma](f))^*\right\|_{L^2_w} \leq C\left\|([b,T_\sigma](f))^{\neq}\right\|_{L^2_w} \\
&\leq C\,\|b\|_{BMO}\left(\|M_\gamma(f)\|_{L^2_w}+\|M_\gamma(T_\sigma f)\|_{L^2_w}\right) \qquad \text{if } 2<\gamma<\infty \\
&\leq C\,\|b\|_{BMO}\left(\|f\|_{L^2_w}+\|T_\sigma f\|_{L^2_w}\right) \qquad \text{if } 1<\gamma<2 \\
&\leq C\,\|b\|_{BMO}\,\|f\|_{L^2_w}
\end{aligned}$$

(cf. Theorem 2.12 in [11]). Because of lemma 1.3, we can extend T_σ to a bounded operator on $L^2(\mathbb{R}^n, wdx)$ and consequently on multiplier spaces $\mathcal{M}\left(H^r \to L^2\right)$.

□

References

[1] Chanillo, S., A note on commutators, *Indiana Univ. Math. J.*, **31** (1982), 7-16.

[2] Cordoba, A., and Fefferman, C., A weighted norm inequality for singular integrals, *Studia Math.* **57** (1976), 97-101.

[3] Coifman, R., and Meyer, Y., Au delà des opé rateurs pseudo-différentiels, Astérisque, **57**, 1978.

[4] Coifman, R., Rocheberg, R., and G. Weiss, Factorization theorems for Hardy spaces in several variables, *Ann. of Math.*, **103** (1976), 611-635.

[5] Fefferman, C., L^p-bounds for pseudo-differential operators, *Israel J. Math.* **14** (1973), 413-417.

[6] Fefferman, C. and Stein, E.M., H^p spaces of several variables, *Acta Math.* **129** (1972), 137-193.

[7] Gala, S., Opérateurs de multiplication ponctuelle entre espace de Sobolev, Ph.D Thesis, University of Evry, France 2005.

[8] Gala, S., Multipliers spaces and pseudo-differential operators, preprint **2005**, France

[9] Garcia-Cuerva, J. and Rubio de Francia, J. L., Weigthed Norm Inequalities and Related Topics. North-Holland, 1985.

[10] Lemarié-Rieusset, P.G., and Gala.S., Multipliers between Sobolev spaces and fractional differentiation, *J. Math. Anal. Appl.*, **322** (2006),1030-1054.

[11] Miller, N., Weighted Sobolev spaces and pseudo-differential operators with smooth symbols, *Trans. Amer. Math. Soc.*, **269** (1982), 91-109.

[12] Stein, E. M., Harmonic Analysis : Real-Variable Methods, Orthogonality, and Oscillatory Integrals, Princeton Univ. Press, Princeton, New Jersey, 1993.

[13] Taylor, M.E., Pseudodifferential Operators and Nonlinear PDE, Birkhauser, Boston, 1991.

[14] Torchinsky, A., Real-Variable Methods in Harmonic Analysis, Academic Press, 1986.

In: African Diaspora Mathematics Research Progress
Editor: Toka Diagana, pp. 55–73
ISBN 978-1-60456-204-0
© 2008 Nova Science Publishers, Inc.

Chapter 4

RBSDEs with Stochastic Monotone and Polynomial Growth Condition

K. Bahlali *
UFR Sciences, Université de Toulon et du Var,
BP 132, 835957 La Garde cedex, France
A. Elouaflin †
UFRMI, Université de Cocody,
22 BP 582 Abidjan, Côte d'Ivoire
M. N'zi ‡
UFRMI, Université de Cocody,
22 BP 582 Abidjan, Côte d'Ivoire

Abstract

In this paper, we are concerned with reflected backward stochastic differential equations (RBSDEs) in a domain of a lower semi-continuous convex function with stochastic monotone and polynomial growth generators. We prove existence and uniqueness result for fixed terminal time. Our work provides an extension of the result established under uniform monotonicity condition.

AMS Subject Classification: Primary 60H20, 60H30.

Keywords: Reflected backward stochastic differential equations, stochastic monotone condition, polynomial growth generators, subdiffential operator.

1 Introduction

It is well known that nonlinear backward stochastic differential equations, in short BSDEs, as formulated in [12], provide probabilistic formulas for solutions of semilinear partial differential equations (PDEs), see for instance Peng [13], Pardoux & Peng[14] and Pardoux

*E-mail: bahlali@univ-tln.fr
†E-mail: elabouo@yahoo.fr
‡E-mail: nzim@ucocody.ci, modestenzi@yahoo.fr

[16, 17]. These equations are an essential mathematical tool in many applications such as mathematical finance (see El Karoui et al. [7]), stochastic optimal control (see Hamadène and Lepeltier [9]). Since the work of Pardoux & Peng [14], many authors have considered BSDEs under weaker conditions than the classical uniform Lipschitz one. Among others, we refer to the work of Darling and Pardoux [5], Briand and Carmona [4], Kobylanski et al.[10] and El Karoui and Huang [6]. In the last paper, the authors introduced the so-called stochastic Lipschitz condition. This means that the Lipschitz coefficients are allowed to be $\mathcal{F}_t-$ adapted processes. The interest of this type of extension comes from the fact that in many applications the usual conditions are not satisfied. Indeed, the Lipschitz coefficient can be an unbounded process which is not deterministic. For example, the pricing of an european claim is equivalent to solve the linear BSDE

$$\begin{cases} -dY_t &= [r(t)Y(t)+\theta(t)Z(t)]dt - Z(t)dW_t \\ Y_t &= \xi \end{cases} \tag{1.1}$$

where ξ is the contingent claim, $r(t)$ is the interest rate and $\theta(t)$ is the risque premium vector. Both $r(t)$ and $\theta(t)$ are not bounded in general. So, it is not possible to solve the equation (1.1) by Pardoux-Peng's Theorem. In order to establish existence and uniqueness result in that framework one requires strong integrability conditions on the data as well as on the solutions.

On the other hand, many papers deal with reflected backward stochastic differential equations, in short RBSDEs(see for instance [8], [15]). These equations are standard BSDE with an additional continuous process added to keep the solution in a fixed domain. This kind of equations appears in finance when one try to find a portfolio to realize an objective at a future time T with the constraint that the wealth of the agent is never negative.

In this paper, we consider RBSDEs with stochastic monotone and polynomial growth generators. We prove an existence and uniqueness result for fixed terminal time. Our work provides an extension of results obtained under uniform monotonicity condition by Briand and Carmona [4] for non reflected BSDEs and by Bahlali et al.[2] in the reflected case. Unlike to the usual monotone case [17], in the stochastic monotone one, it is not possible to choose the monotonicity constant (process) in y to be vanished. This is due to the fact that the right space which contains the solution is related to the processes appearing in the monotonicity condition.

The paper is organized as follows. Notations, assumptions and definitions are stated in section 2. Section 3 treats the existence and uniqueness without reflection and section 4 deals with the reflected case.

2 Notations, Assumptions and Definitions

2.1 Notations

Let $W = \{W_t, \mathcal{F}_t, t \geq 0\}$ be a n -dimensional Brownian motion defined on a complete probability space $(\Omega, \mathcal{F}, \mathbb{P})$. $\{\mathcal{F}_t, t \geq 0\}$ stands for the natural filtration of W, augmented with the $\mathbb{P}$ -nul sets of $\mathcal{F}$. The inner product of $\mathbb{R}^d$ is denoted by $\langle , \rangle$ and the Euclidean norm by $|.|$. For every $z \in \mathbb{R}^{d\times n}$ we put $|Z|^2 = tr(ZZ^*)$.

For every $\mathcal{F}_t-$adapted process a with positive values, we define an increasing process A by

setting $A(t) = \int_0^t a^2(s)ds$.
For every $\beta > 0$ and a every terminal time T, we consider the following spaces:

$$L^2(\beta, a, T, \mathbb{R}^d) = \left\{ \begin{array}{l} \xi; \mathbb{R}^d\text{-valued, } \mathcal{F}_T-\text{measurable random variables} \\ \text{such that } \|\xi\|_\beta^2 = \mathbb{E}\left(e^{\beta A(T)}|\xi|^2\right) < +\infty. \end{array} \right\}$$

$$L^2(\beta, a, [0,T], \mathbb{R}^d) = \left\{ \begin{array}{l} Y; \mathbb{R}^d\text{-valued, } \mathcal{F}_t-\text{adapted processes such that} \\ \|Y\|_\beta^2 = \mathbb{E}\left(\int_0^T e^{\beta A(s)}|Y(s)|^2 ds\right) < +\infty. \end{array} \right\}$$

$$L^{2,a}(\beta, a, [0,T], \mathbb{R}^d) = \left\{ \begin{array}{l} Y; \mathbb{R}^d\text{-valued, } \mathcal{F}_t-\text{adapted processes such that} \\ \|aY\|_\beta^2 = \mathbb{E}\left(\int_0^T e^{\beta A(s)}a^2(s)|Y(s)|^2 ds\right) < +\infty. \end{array} \right\}$$

$$L_c^2(\beta, a, [0,T], \mathbb{R}^d) = \left\{ \begin{array}{l} Y; \mathbb{R}^d\text{-valued, continuous } \mathcal{F}_t-\text{adapted processes} \\ \text{such that } \|Y\|_{\beta,c}^2 = \mathbb{E}\left(\sup_{0\le s\le T} e^{\beta A(s)}|Y(s)|^2\right) < +\infty. \end{array} \right\}$$

$L^2(\beta, a, [0,T], \mathbb{R}^d)$ endowed with the norm $\|.\|_\beta$ is a Banach space. Therefore, so is the space

$$\mathcal{M}(\beta, a, T) = L^{2,a}(\beta, a, [0,T], \mathbb{R}^d) \times L^2(\beta, a, [0,T], \mathbb{R}^{d\times n})$$

endowed with the norm $\|(Y,Z)\|_\beta^2 = \|aY\|_\beta^2 + \|Z\|_\beta^2$.

We denote by $\mathcal{M}_c(\beta, a, T)$ the subspace of $\mathcal{M}(\beta, a, T)$ defined by

$$\mathcal{M}^c(\beta, a, T) = \left(L^{2,a}(\beta, a, [0,T], \mathbb{R}^d) \cap L^{2,c}(\beta, a, [0,T], \mathbb{R}^d)\right) \times L^2(\beta, a, [0,T], \mathbb{R}^{d\times n});$$

and endowed with the norm $\|(Y,Z)\|_{\beta,c}^2 = \|Y\|_{\beta,c}^2 + \|aY\|_\beta^2 + \|Z\|_\beta^2$.

For every $p > 1$, we denote by $\mathcal{M}_p^c(\beta, a, T)$ the space of couple of processes (Y,Z) in $\mathcal{M}^c(\beta, a, T)$ such that

$$\mathbb{E}\left(\sup_{0\le s\le T} e^{p\beta A(s)}|Y(s)|^{2p} + \left(\int_0^T e^{\beta A(s)}a^2(s)|Y(s)|^2 ds\right)^p + \left(\int_0^T e^{\beta A(s)}|Z(s)|^2 ds\right)^p\right) < +\infty.$$

Remark 2.1. If a and b are two $\mathcal{F}_t-$adapted processes with positive values such that $b > a$ then $L^2(\beta, b, [0,T], \mathbb{R}^d) \subset L^2(\beta, a, [0,T], \mathbb{R}^d)$. Therefore, $\mathcal{M}^c(\beta, b, T) \subset \mathcal{M}^c(\beta, a, T)$.

2.2 Assumptions and Definitions

Let $f : \Omega \times [0,T] \times \mathbb{R}^d \times \mathbb{R}^{d\times n} \longrightarrow \mathbb{R}^d$ be a function such that for all $(y,z) \in \mathbb{R}^d \times \mathbb{R}^{d\times n}$, $f(.,.,y,z)$ is progressively measurable, and ξ an $\mathbb{R}^d-$valued $\mathcal{F}_T-$measurable random variable.
For some $\beta > 0$, we assume that the triple (T, ξ, f) satisfies the conditions:

(H1) There exist a $\mathcal{F}_t-$adapted process $\theta(t)$ and a nonnegative $\mathcal{F}_t-$adapted process $\nu(t)$ such that $\forall (y, y', z, z') \in \mathbb{R}^d \times \mathbb{R}^d \times \mathbb{R}^{d\times n} \times \mathbb{R}^{d\times n}$,

$$\left\{ \begin{array}{ll} (i) & \langle y - y', f(t,y,z) - f(t,y',z) \rangle \le \theta(t)|y-y'|^2 \\ (ii) & |f(t,y,z) - f(t,y,z')| \le \nu(t)|z-z'| \\ (iii) & y \mapsto f(.,.,y,z) \text{ is continuous } dt \otimes d\mathbb{P} \text{ a.s.} \end{array} \right.$$

(H2) There exist $\varepsilon > 0$ such that $a^2(t) = |\theta(t)| + v^2(t) > \varepsilon$.
If **(H2)** is not fulfilled, replace $v(t)$ by $v(t) + \sqrt{\varepsilon}$.

(H3) There exist a $\mathcal{F}_t-$adapted process η with values in $[0,+\infty]$, $p > 1$ and a positive constant K such that

$$(iv) \quad \mathbb{E}\left[\left(\int_0^T e^{\beta A(s)} |\eta(s)|^2 ds\right)^p\right] < +\infty,$$

$$(v) \quad |f(t,y,z)| \leq |f(t,0,z)| + \eta(t) + K(1+|y|^p),$$

$$(vi) \quad \mathbb{E}\left[e^{p\beta A(T)}\left(1+|\xi|^{2p}\right)\right] < +\infty,$$

$$(vii) \quad \mathbb{E}\left[\left(\int_0^T e^{\beta A(s)} \frac{|f(.,0,0))|^2}{a^2(s)} ds\right)^p\right] < +\infty.$$

Remark 2.2. Let us note that the case $p = 1$ has been treated by Bahlali et al. [1].

Definition 2.3. A solution of the BSDE with data (T,ξ,f) is a pair of $\mathcal{F}_t-$adapted processes (Y,Z) with values in $\mathbb{R}^d \times \mathbb{R}^{d\times n}$ such that
(j) $(Y,Z) \in \mathcal{M}_p^c(\beta,a,T)$,
and
(jj)

$$Y_t = \xi + \int_t^T f(s,Y_s,Z_s)ds - \int_t^T Z_s dW_s.$$

3 Existence and Uniqueness Result for BSDEs

3.1 Uniqueness

We first state a priori estimates on the solutions.

Proposition 3.1. *Assume that the data* (T,ξ,f), $\left(T,\xi',f'\right)$ *satisfy* **(H1)**-**(H3)** *with the same coefficients*[1]. *Let* (Y,Z) *(resp.* (Y',Z')*) be a solution of the corresponding BSDE and put* $\Delta f(t) = f(t,Y_t',Z_t') - f'(t,Y_t',Z_t')$, $\Delta Y_t = Y_t - Y_t'$, $\Delta Z_t = Z_t - Z_t'$, $\Delta\xi = \xi - \xi'$.
If $\mathbb{E}\left(\int_0^T e^{\beta A(s)} \frac{|\Delta f(s)|^2}{a^2(s)} ds\right)^p < +\infty$, *then for* β *sufficiently large, there exists a constant* $C(p)$ *which depends only on p such that*
(i)

$$\mathbb{E}\left(\sup_{0\leq t\leq T} e^{p\beta A(t)} |\Delta Y_t|^{2p}\right) \leq C(p)\Gamma$$

(ii)

$$\mathbb{E}\left[\left(\int_0^T e^{\beta A(s)} |\Delta Z_s|^2 ds\right)^p + \left(\int_0^T e^{\beta A(s)} a^2(s) |\Delta Y_s|^2 ds\right)^p\right] \leq C(p)\Gamma$$

[1]This assumption is not restrictive since we may choose $a^2(t) = \overline{\theta}(t) + \overline{v}^2(t)$, with $\overline{\theta}(t) = \max(\theta(t),\theta'(t)), \overline{v}(t) = \max(v(t),v'(t))$

where

$$\Gamma = \mathbb{E}\left[(e^{p\beta A(T)} |\Delta\xi|^{2p}) + \frac{1}{\beta^p} \left(\int_0^T e^{\beta A(s)} \frac{|\Delta f(s)|^2}{a^2(s)} ds \right)^p \right].$$

Proof. In the sequel, $C(p)$ denotes a constant depending only on p which may vary from line to line.
By virtue of Itô's formula, we have

$$\begin{aligned} & e^{\beta A(t)} |\Delta Y_t|^2 + \beta \int_t^T e^{\beta A(s)} a^2(s) |\Delta Y_s|^2 ds + \int_t^T e^{\beta A(s)} |\Delta Z_s|^2 ds \\ = \; & e^{\beta A(T)} |\Delta\xi|^2 + 2 \int_t^T e^{\beta A(s)} \langle \Delta Y_s, f(s, Y_s, Z_s) - f'(s, Y'_s, Z'_s) \rangle ds \\ & -2 \int_t^T e^{\beta A(s)} \langle \Delta Y_s, \Delta Z_s dW_s \rangle. \end{aligned} \tag{3.1}$$

In view of **(H1-i)**, **(H1-ii)** and Young's inequalities $2ab \le ka^2 + \frac{1}{k} b^2$, for every $k > 0$ we have

$$\begin{aligned} & 2\langle \Delta Y_s, f(s, Y_s, Z_s) - f'(s, Y'_s, Z'_s) \rangle \\ \le \; & 2\theta(s) |\Delta Y_s|^2 + 2 |\Delta Y_s| [\nu(s) |\Delta Z_s| + |\Delta f(s)|] \\ \le \; & \left(\frac{\beta}{2} + 2 \right) a^2(s) |\Delta Y_s|^2 + \frac{1}{2} |\Delta Z_s|^2 + \frac{2}{\beta} \frac{|\Delta f(s)|^2}{a^2(s)}. \end{aligned}$$

Therefore, by virtue of (3.1) we deduce that

$$\begin{aligned} & e^{\beta A(t)} |\Delta Y_t|^2 + \left(\frac{\beta}{2} - 2 \right) \int_t^T e^{\beta A(s)} a^2(s) |\Delta Y_s|^2 ds + \frac{1}{2} \int_t^T e^{\beta A(s)} |\Delta Z_s|^2 ds \\ \le \; & e^{\beta A(T)} |\Delta\xi|^2 + \frac{2}{\beta} \int_t^T e^{\beta A(s)} \frac{|\Delta f(s)|^2}{a^2(s)} ds - 2 \int_t^T e^{\beta A(s)} \langle \Delta Y_s, \Delta Z_s dW_s. \rangle \end{aligned} \tag{3.2}$$

By Choosing $\beta > 4$, and by taking the conditional expectation with respect to $\mathcal{F}_t$, we obtain

$$e^{\beta A(t)} |\Delta Y_t|^2 \le \mathbb{E}\left(e^{\beta A(T)} |\Delta\xi|^2 + \frac{2}{\beta} \int_0^T e^{\beta A(s)} \frac{|\Delta f(s)|^2}{a^2(s)} ds \,|\, \mathcal{F}_t \right).$$

Since $p > 1$, Doob's maximal inequality yields

$$\mathbb{E}\left(\sup_{0\le t\le T} e^{p\beta A(t)} |\Delta Y_t|^{2p} \right) \le C(p) \mathbb{E}\left[e^{p\beta A(T)} |\Delta\xi|^{2p} + \frac{1}{\beta^p} \left(\int_0^T e^{\beta A(s)} \frac{|\Delta f(s)|^2}{a^2(s)} ds \right)^p \right].$$

So, we have proved (i).
Let us deal with the proof of (ii). In view of (3.2), we derive

$$\begin{aligned} & \sup_{0\le t\le T} e^{\beta A(t)} |\Delta Y_t|^2 + \left(\frac{\beta}{2} - 2 \right) \int_0^T e^{\beta A(s)} a^2(s) |\Delta Y_s|^2 ds + \frac{1}{2} \int_0^T e^{\beta A(s)} |\Delta Z_s|^2 ds \\ \le \; & e^{\beta A(T)} |\Delta\xi|^2 + \frac{2}{\beta} \int_0^T e^{\beta A(s)} \frac{|\Delta f(s)|^2}{a^2(s)} ds + 2 \sup_{0\le t\le T} \left| \int_t^T e^{\beta A(s)} \langle \Delta Y_s, \Delta Z_s dW_s \rangle \right|. \end{aligned}$$

By virtue of Burkhölder-Davis-Gundy's inequality and $ab \leq \frac{a^2}{2} + \frac{b^2}{2}$, we deduce that

$$\begin{aligned}
&\mathbb{E}\left(\int_0^T e^{\beta A(s)}|\Delta Z_s|^2 ds\right)^p \\
\leq\ & C(p)\Gamma + C(p)\mathbb{E}\left(\int_0^T e^{2\beta A(s)}|\Delta Y_s|^2|\Delta Z_s|^2 ds\right)^{\frac{p}{2}} \\
\leq\ & C(p)\Gamma + C(p)\mathbb{E}\left(\sup_{0\leq t\leq T} e^{\frac{p}{2}\beta A(t)}|\Delta Y_t|^p\left(\int_0^T e^{\beta A(s)}|\Delta Z_s|^2 ds\right)^{\frac{p}{2}}\right) \\
\leq\ & C(p)\Gamma + C(p)\mathbb{E}(\sup_{0\leq t\leq T} e^{p\beta A(t)}|\Delta Y_t|^{2p}) + \frac{1}{2}\mathbb{E}\left(\int_0^T e^{\beta A(s)}|\Delta Z_s|^2 ds\right)^p.
\end{aligned}$$

So, thanks to (i), we obtain

$$\mathbb{E}\left(\int_0^T e^{\beta A(s)}|\Delta Z_s|^2 ds\right)^p \leq C(p)\Gamma.$$

Similarly, we show that

$$\mathbb{E}\left(\int_0^T e^{\beta A(s)}a^2(s)|\Delta Y_s|^2 ds\right)^p \leq C(p)\Gamma.$$

By combining these inequalities, we derive (ii). □

Proposition 3.2. *Let (Y,Z) be a solution of the BSDE with data (T,ξ,f). Then*

$$\begin{aligned}
&\mathbb{E}\left[\sup_{0\leq t\leq T} e^{p\beta A(t)}|Y_t|^{2p} + \left(\int_0^T e^{\beta A(s)}|Z_s|^2 ds\right)^p + \left(\int_0^T e^{\beta A(s)}a^2(s)|Y_s|^2 ds\right)^p\right] \\
\leq\ & C(p)\mathbb{E}\left[(e^{p\beta A(T)}|\xi|^{2p}) + \frac{1}{\beta^p}\left(\int_0^T e^{\beta A(s)}\frac{|f(s,0,0)|^2}{a^2(s)}ds\right)^p\right].
\end{aligned}$$

Proof. By virtue of Itô's formula, we have

$$\begin{aligned}
&e^{\beta A(t)}|Y_t|^2 + \beta\int_t^T e^{\beta A(s)}a^2(s)|Y_s|^2 ds + \int_t^T e^{\beta A(s)}|Z_s|^2 ds \\
=\ & e^{\beta A(T)}|\xi|^2 + 2\int_t^T e^{\beta A(s)}\langle Y_s, f(s,Y_s,Z_s)\rangle ds - 2\int_t^T e^{\beta A(s)}\langle Y_s, Z_s dW_s\rangle.
\end{aligned}$$

Now,

$$2\langle Y_s, f(s,Y_s,Z_s)\rangle \leq \left(\frac{\beta}{2}+2\right)a^2(s)|Y_s|^2 + \frac{1}{2}|Z_s|^2 + \frac{2}{\beta}\frac{|f(s,0,0)|^2}{a^2(s)}.$$

Therefore,

$$\begin{aligned}
&e^{\beta A(t)}|Y_t|^2 + \left(\frac{\beta}{2}-2\right)\int_t^T e^{\beta A(s)}a^2(s)|Y_s|^2 ds + \frac{1}{2}\int_t^T e^{\beta A(s)}|Z_s|^2 ds \\
\leq\ & e^{\beta A(T)}|\xi|^2 + \frac{2}{\beta}\int_t^T e^{\beta A(s)}\frac{|f(s,0,0)|^2}{a^2(s)}ds - 2\int_t^T e^{\beta A(s)}\langle Y_s, Z_s dW_s\rangle.
\end{aligned}$$

Now, the rest of the proof follows similar arguments in the proof of Proposition 3.1 so it is omitted. □

Corollary 3.3. *Assume* **(H1)–(H3)**. *Then, the BSDE* $(j)-(jj)$ *(definition 2.3) has at most one solution.*

Proof. It's an immediate consequence of Proposition 3.1. □

3.2 Existence

The main result of this section is the following.

Theorem 3.4. *Assume* **(H1)–(H3)**. *Then, for* β *sufficiently large, the BSDE* $(j)-(jj)$ *(definition 2.3) has at least one solution.*

The proof is based on a fixed point argument and uses a priori estimates and the following

Proposition 3.5. *Assume* **(H1) - (H3)**. *Let* $\{V_t : 0 \le t \le T\}$ *be a* $\mathcal{F}_t$*-adapted process satisfying* $\mathbb{E}\left(\int_0^T e^{\beta A(s)} |V_s|^2 ds\right)^p < +\infty$. *Then there exists a couple of* $\mathcal{F}_t$*-adapted processes* (Y,Z) *with values in* $\mathbb{R}^d \times \mathbb{R}^{d\times n}$ *such that*
(3j)

$$\mathbb{E}\left[\sup_{0\le t\le T} e^{p\beta A(t)}|Y_t|^{2p} + \left(\int_0^T e^{\beta A(s)}|Z_s|^2 ds\right)^p + \left(\int_0^T e^{\beta A(s)} a^2(s)|Y_s|^2 ds\right)^p\right] < +\infty$$

and
(4j)

$$Y_t = \xi + \int_t^T f(s,Y_s,V_s)ds - \int_t^T Z_s dW_s.$$

Proof. In the sequel, we put $h(s,y) = f(s,y,V_s)$ for every $s \in [0,T]$ and we split the proof in two steps.
Step 1. We set $\overline{\xi} = e^{\frac{\beta}{2}A(T)}|\xi|$ and assume that

$$|\overline{\xi}|^2 + \sup_{0\le t\le T} |h(t,0)|^2 \le C. \tag{3.3}$$

Let φ_q be a smooth function from $\mathbb{R}^d$ to $\mathbb{R}_+$ such that $0 \le \varphi_q \le 1$; $\varphi_q(x) = 1$ if $|x| \le q$ and $\varphi_q(x) = 0$ if $|x| \ge q+1$. For every $n \in \mathbb{N}^*$, we put $q(n) = [(C + \frac{6T}{\beta\varepsilon}(nC + n^3 + 4K^2 n))^{\frac{1}{2}}]$, where $[r]$ is the integer part of r.
Let us define

$$h_n(t,y) = h_{n,\,q(n)+2}(t,y) = \mathbf{1}_{\{\eta(t)+e^{\beta A(t)}\le n\}} \int_{\mathbb{R}^d} \varphi_{q(n)+2}(y-u)h(t,y-u)\rho_n(u)du,$$

where $\rho_n : \mathbb{R}^d \longrightarrow \mathbb{R}_+$ is a sequence of smooth functions with compact support in the ball $B(0,1)$ which approximates the Dirac measure at 0 and satisfies $\int_{\mathbb{R}^d} \rho_n(u)du = 1$, for every $n \in \mathbb{N}^*$.
It is obvious that $\{h_n(t,.) : n \in \mathbb{N}^*\}$ is a sequence of smooth functions with compact support satisfying: for every $t \in [0,T]$

(**a**) $h_n(t,.)$ converges towards $h(t,.)$ on compact sets as $n \to +\infty$, (**b**) for every $n \in \mathbb{N}^*$, $h_n(t,.)$ is globally stochastic Lipschitz with coefficient $K(n,t) = a^2(t) + C_n$, where

$$C_n = \frac{1}{4}\alpha_n^2 \frac{C}{\varepsilon} + \alpha_n \left[n + K\left(2 + (q(n)+3)^p\right)\right] \text{ and } \alpha_n = \int_{\mathbb{R}^d} |\Delta\rho_n(u)|du$$

(**c**) For every y, y' such that

$$max(|y|, |y'|) \leq q(n)+1,\ \langle y - y', h_n(t,y) - h_n(t,y')\rangle \leq \mathbf{1}_{\{\eta(t)+e^{\beta A(t)} \leq n\}} \theta(t)|y-y'|^2.$$

(**d**) $|h_n(t,y)| \leq \mathbf{1}_{\{\eta(t)+e^{\beta A(t)} \leq n\}} \left(|h(t,0)| + \eta(t) + K(2+|y|^p)\right)$.

Now, let us put

$$a_n^2(t) = K(n,t),$$

and

$$A_n(t) := \int_0^t a_n^2(s)ds = A(t) + C_n t.$$

In view of [6], the equation

$$Y_t^n = \xi + \int_t^T h_n(s, Y_s^n)ds - \int_t^T Z_s^n dW_s \tag{3.4}$$

has a unique solution (Y^n, Z^n) which belongs to the space $\mathcal{M}_c(\beta, a_n, T)$.
By virtue of Remark 2.1, we have $(Y^n, Z^n) \in \mathcal{M}_c(\beta, a, T)$.
Moreover, (3.3) implies that for every $t \in [0, T]$

$$|Y_t^n|^2 \leq C + \frac{6T}{\beta\varepsilon}(nC + n^3 + 4K^2 n), \tag{3.5}$$

which justifies the choice of the integer $q(n)$.
Now, we aim to show that under (3.3), the sequence $\{(Y^n, Z^n) : n \in \mathbb{N}^*\}$ converges towards the solution of the BSDE $(4j)$ as $n \longrightarrow +\infty$. To this end, we prove two intermediate results.

Lemma 3.6. *Assume* **(H1)** - **(H3)** *and (3.3). Then, for* β *sufficiently large, there exist a constant* C *depending only on* p, β, K *and* ε *such that*

$$(i) \quad \sup_{n\in\mathbb{N}^*} \mathbb{E}\left\{ \sup_{0\leq t\leq T} e^{p\beta A(t)} |Y_t^n|^{2p} + \left(\int_0^T e^{\beta A(s)} a^2(s) |Y_s^n|^2 ds\right)^p + \left(\int_0^T e^{\beta A(s)} |Z_s^n|^2 ds\right)^p \right\}$$
$$\leq C\mathbb{E}\left\{ e^{p\beta A(T)}(1+|\xi|^{2p}) + \left(\int_0^T e^{\beta A(s)} \eta(s)^2 ds\right)^p \right\}.$$

and

$$(ii) \quad \sup_{n\in\mathbb{N}^*} \mathbb{E}\left\{ \int_0^T e^{\beta A(s)} |h_n(s, Y_s^n)|^2 ds \right\} < +\infty.$$

Proof. (i) By virtue of Itô's formula, we have

$$\begin{aligned} & e^{\beta A(t)}\,|Y_t^n|^2+\beta\int_t^T e^{\beta A(s)}a^2(s)\,|Y_s^n|^2\,ds+\int_t^T e^{\beta A(s)}\,|Z_s^n|^2\,ds \\ = \;& e^{\beta A(T)}\,|\xi|^2+2\int_t^T e^{\beta A(s)}\,\langle Y_s^n,h_n(s,Y_s^n)\rangle\,ds-2\int_t^T e^{\beta A(s)}\,\langle Y_s^n,Z_s^n dW_s\rangle . \end{aligned} \tag{3.6}$$

Since

$$\begin{aligned} \langle Y_s^n,h_n(s,Y_s^n)\rangle &= \langle Y_s^n,h_n(s,Y_s^n)-h_n(s,0)\rangle+\langle Y_s^n,h_n(s,0)\rangle \\ &\le \left(2+\frac{\beta}{2}\right)a^2(s)\,|Y_s^n|^2+\frac{2}{\beta\varepsilon}\,|h_n(s,0)|^2, \end{aligned}$$

Proposition 3.2 and (3.6) imply that

$$\begin{aligned} & \mathbb{E}\left\{\sup_{0\le t\le T} e^{p\beta A(t)}\,|Y_t^n|^{2p}+\left(\int_0^T e^{\beta A(s)}a^2(s)\,|Y_s^n|^2\,ds\right)^p+\left(\int_0^T e^{\beta A(s)}\,|Z_s^n|^2\,ds\right)^p\right\} \\ \le\; & C(p)\mathbb{E}\left\{e^{p\beta A(T)}\,|\xi|^{2p}+\frac{1}{(\beta\varepsilon)^p}\left(\int_0^T e^{\beta A(s)}\,|h_n(s,0)|^2\,ds\right)^p\right\}. \end{aligned}$$

Therefore, by virtue of **(d)** we deduce (i).
Now, it suffices to use (i) and **(H3)** to derive (ii). □

Lemma 3.7. *Assume* **(H1)**–**(H3)** *and (3.3).Then* $\{(Y^n,Z^n):n\in\mathbb{N}^*\}$ *is a Cauchy sequence in* $\mathcal{M}_c(\beta,a,T)$.

Proof. Let $m\ge n$, and put $\Delta Y_t=Y_t^m-Y_t^n$, $\Delta Z_t=Z_t^m-Z_t^n$.
By virtue of Itô's formula, Burkhölder-Davis-Gundy's inequality and **(c)**, we derive that

$$\begin{aligned} & \mathbb{E}\left(\sup_{0\le t\le T} e^{\beta A(t)}\,|\Delta Y_t|^2\right)+(2\beta-4)\mathbb{E}\left(\int_0^T e^{\beta A(s)}a^2(s)\,|\Delta Y_s|^2\,ds\right) \\ & +\mathbb{E}\left(\int_0^T e^{\beta A(s)}\,|\Delta Z_s|^2\,ds\right) \\ \le\; & C\mathbb{E}\left(\int_0^T e^{\beta A(s)}\,|\Delta Y_s|\,|h_m(s,Y_s^n)-h_n(s,Y_s^n)|\,ds\right). \end{aligned} \tag{3.7}$$

For a every $M>0$, we set

$$B_{m,n}^M:=\{(s,\omega)\in[0,T]\times\Omega;\,|Y_s^m|+|Y_s^n|>M\},\;\overline{B}_{m,n}^M:=[0,T]\times\Omega\setminus B_{m,n}^M$$

and put

$$\begin{aligned} I_{m,n} &= \mathbb{E}\left(\int_0^T e^{\beta A(s)}\,|\Delta Y_s|\,|h_m(s,Y_s^n)-h_n(s,Y_s^n)|\,ds\right) \\ I_{m,n}^1 &= \mathbb{E}\left(\int_0^T e^{\beta A(s)}\,|\Delta Y_s|\,|h_m(s,Y_s^n)-h_n(s,Y_s^n)|\,\mathbf{1}_{\overline{B}_{m,n}^M}\,ds\right) \\ I_{m,n}^2 &= \mathbb{E}\left(\int_0^T e^{\beta A(s)}\,|\Delta Y_s|\,|h_m(s,Y_s^n)-h_n(s,Y_s^n)|\,\mathbf{1}_{B_{m,n}^M}\,ds\right). \end{aligned}$$

By virtue of Holder's inequality and Young's inequality, we have

$$\begin{aligned} I^2_{m,n} &\leq \frac{1}{M^{p-1}}\mathbb{E}\left\{\int_0^T e^{\beta A(s)}\,|\Delta Y_s|\,|h_m(s,Y^n_s)-h_n(s,Y^n_s)|\,(|Y^m_s|+|Y^n_s|)^{p-1}\,ds\right\} \\ &\leq \frac{C(T,p)}{M^{p-1}}\sup_{n\in N^*}\mathbb{E}\left(\sup_{0\leq t\leq T} e^{\beta A(t)}\,|Y^n_t|^{2p}\right) \\ &\quad +\frac{C(T,p)}{M^{p-1}}\sup_{n\in N^*}\mathbb{E}\left(\int_0^T e^{\beta A(s)}\,|h_m(s,Y^n_s)-h_n(s,Y^n_s)|^2\,ds\right). \end{aligned}$$

Now, in view of Lemma 3.6, we deduce that

$$I^2_{m,n} \leq \frac{C(T,p)}{M^{p-1}}.$$

Since M is arbitrary, $I^2_{m,n}$ can be made arbitrary small by choosing M large enough. Now, let us treat $I^1_{m,n}$. We have

$$\begin{aligned} I^1_{m,n} &\leq M\mathbb{E}\left(\int_0^T e^{\beta A(s)}\,|h_m(s,Y^n_s)-h_n(s,Y^n_s)|\,1_{\{|Y^n_s|\leq M\}}ds\right) \qquad (3.8)\\ &\leq M\mathbb{E}\left(\int_0^T \sup_{|y|\leq M} e^{\beta A(s)}\,|h_m(s,y)-h_n(s,y)|\,ds\right). \end{aligned}$$

Since $h_n(t,.)$ converges towards $h(t,.)$ on compacts sets and $\sup_{|y|\leq M} e^{\beta A(s)}\,|h_n(s,y)| \leq \{|h(s,0)|+\eta(s)+K(2+M^p)\}\,e^{\beta A(s)}$, Lebesgue convergence theorem ensures that the right hand side of (3.8) tends to zero as $m,n \longrightarrow +\infty$.
Hence, in view of (3.7), we conclude that

$$\begin{aligned} &\mathbb{E}\left(\sup_{0\leq t\leq T} e^{\beta A(t)}\,|\Delta Y_t|^2\right)+(2\beta-4)\mathbb{E}\left(\int_0^T e^{\beta A(s)}a^2(s)\,|\Delta Y_s|^2\,ds\right) \\ &+\mathbb{E}\left(\int_0^T e^{\beta A(s)}\,|\Delta Z_s|^2\,ds\right) \end{aligned}$$

tends to zero as $m,n \longrightarrow +\infty$. □

which means that $\{(Y^n,Z^n) : n\in\mathbb{N}^*\}$ is a Cauchy sequence in the Banach space $\mathcal{M}_c(\beta,a,T)$. Therefore it converges to a process (Y,Z). Now, by virtue of Lemma 3.6 the sequence $\{(Y^n,Z^n) : n\in\mathbb{N}^*\}$ is uniformly bounded in $\mathcal{M}^p_c(\beta,a,T)$. Therefore, Fatou's lemma implies that (Y,Z) belongs to the space $\mathcal{M}^p_c(\beta,a,T)$. Finally, passing to limit in (3.4) one can show that the process (Y,Z) solves the equation $(4j)$.

Step 2

Let

$$\xi_n = \frac{\inf(n, e^{\frac{\beta A(T)}{2}}\,|\xi|)}{e^{\frac{\beta A(T)}{2}}\,|\xi|}\xi,$$

and

$$h_n(t,y)=\begin{cases} h(t,y)-h(t,0)+\frac{\inf(n,|h(t,0)|)}{|h(t,0)|}h(t,0) & \text{if } h(t,0)\neq 0.\\ h(t,y) & \text{if } h(t,0)=0.\end{cases}$$

Let us note that $\mathbb{E}\left(e^{\beta A(T)}|\xi_n-\xi|^2\right)$ and $\mathbb{E}\left(\int_0^T e^{\beta A(s)}\frac{|h_n(s,0)-h(s,0)|^2}{a^2(s)}ds\right)$ tend to zero as $n\longrightarrow 0$.
Since (ξ_n,h_n) satisfies (3.3), for each $n\in\mathbb{N}^*$, there exists an unique process (Y^n, Z^n) which satisfy $(3j)$ and

$$Y_t^n=\xi_n+\int_t^T h_n(s,Y_s^n)ds-\int_t^T Z_s^n dW_s,\ 0\le t\le T.$$

By now classical arguments, we have for every $n,m\in\mathbb{N}^*$

$$\begin{aligned}
&\mathbb{E}\left(\sup_{0\le t\le T} e^{\beta A(t)}|Y_t^n-Y_t^m|^2+\left(\frac{\beta}{2}-2\right)\int_0^T e^{\beta A(s)}a^2(s)|Y_s^n-Y_s^m|^2ds\right)\\
&+\mathbb{E}\left(\int_0^T e^{\beta A(s)}|Z_s^n-Z_s^m|^2ds\right)\\
\le\ & C\mathbb{E}\left(e^{\beta A(T)}|\xi_n-\xi_m|^2+\int_0^T e^{\beta A(s)}\frac{|h_n(s,0)-h_m(s,0)|^2}{a^2(s)}ds\right).
\end{aligned}$$

Since the right hand side converges to zero as $n,m\longrightarrow+\infty$, there exists a pair of $\mathcal{F}_t-$adapted processes (Y,Z) such that

$$\lim_n\|(Y^n,Z^n)-(Y,Z)\|_{\beta,c}^2=0$$

and satisfying $(3j)$ and $(4j)$.
Now, the sequence $\{(Y^n,Z^n):n\in\mathbb{N}^*\}$ is uniformly bounded in $\mathcal{M}_c^p(\beta,a,T)$ and Fatou's lemma ensures that (Y,Z) belongs to the space $\mathcal{M}_p^c(\beta,a,T)$. The proof of Proposition 3.5 is now complete. □

Proof of Theorem 3.4. For a every $(U,V)\in\mathcal{M}_p(\beta,a,T)$, by virtue of Proposition 3.5, the BSDE

$$Y_t=\xi+\int_t^T f(s,Y_s,V_s)ds-\int_t^T Z_s dW_s.$$

has an unique solution. So, we can define the mapping

$$\begin{array}{rccc}\Pi: & \mathcal{M}_p(\beta,a,T) & \longrightarrow & \mathcal{M}_p^c(\beta,a,T)\subset\mathcal{M}_p(\beta,a,T)\\ & (U,V) & \longmapsto & \Pi(U,V)\end{array}$$

such that $\Pi(U,V)$ is the unique solution of the corresponding BSDE.
Let $(U,V,U',V')\in\mathcal{M}_p(\beta,a,T)\times\mathcal{M}_p(\beta,a,T)$ and let us put $\Pi(U,V)=(Y,Z),\Pi(U',V')=(Y',Z'),\ \Delta Y_t=Y_t-Y_t',\ \Delta Z_t=Z_t-Z_t'$.
In view of Proposition 3.1, we have

$$\begin{aligned}
&\mathbb{E}\left[\left(\int_0^T e^{\beta A(s)}|\Delta Z_s|^2ds\right)^p+\left(\int_0^T e^{\beta A(s)}a^2(s)|\Delta Y_s|^2ds\right)^p\right]\\
\le\ &\frac{C(p)}{\beta^p}\mathbb{E}\left(\int_0^T e^{\beta A(s)}\frac{\left|f(s,Y_s',V_s)-f(s,Y_s',V_s')\right|^2}{a^2(s)}ds\right)^p.
\end{aligned}$$

By virtue of **(H1-ii)**, we deduce that

$$\begin{aligned}
&\mathbb{E}\left[\left(\int_0^T e^{\beta A(s)}\,|\Delta Z_s|^2\,ds\right)^p + \left(\int_0^T e^{\beta A(s)} a^2(s)\,|\Delta Y_s|^2\,ds\right)^p\right] \\
\leq\; & \frac{C(p)}{\beta^p}\mathbb{E}\left(\int_0^T e^{\beta A(s)}\left|V_s - V_s'\right|^2 ds\right)^p .
\end{aligned}$$

Therefore if β is sufficiently large, Π is a contracting mapping and its unique fixed point solves our BSDE.

4 Reflected Case

Let ξ be a $\mathcal{F}_T$-measurable random variable and $\Phi : \mathbb{R}^d \longrightarrow (-\infty, +\infty]$ be a proper, lower semi-continuous and convex function.
$\partial\Phi$ stands for the sub-differential operator of Φ.
We set

$$\begin{aligned}
Dom(\Phi) &= \left\{x \in \mathbb{R}^d : \; \Phi(x) < +\infty\right\}, \\
\partial\Phi(u) &= \left\{u^* \in \mathbb{R}^d : \; \Phi(z) \geq \Phi(u) + \langle\, z - u,\; u^*\rangle \;, \forall z \in \mathbb{R}^d\right\}, \\
Dom(\partial\Phi) &= \left\{u \in \mathbb{R}^d : \partial\Phi(u) \neq \emptyset\right\}, \\
Gr(\partial\Phi) &= \left\{\; (u, u^*) \in \mathbb{R}^d \times \mathbb{R}^d : u \in Dom(\partial\Phi) \text{ and } u^* \in \partial\Phi(u)\right\}.
\end{aligned}$$

we assume that the triple $(T, \xi, f\,)$ satisfies **(H1)** to **(H3)** for some $\beta > 0$. We also assume that

(H4) $\xi \in \overline{Dom(\Phi)}$, $\mathbb{E}\left(e^{\beta A(T)}\Phi(\xi)\right) < +\infty$.

Now, we introduce our reflected backward stochastic differential equation associated to the data (T, ξ, f, Φ).

Definition 4.1. A solution of the RBSDE (T, ξ, f, Φ) is a triple $\{(Y_t, Z_t, K_t) : 0 \leq t \leq T\}$ of $\mathcal{F}_t$-adapted processes taking values respectively in $\mathbb{R}^d, \mathbb{R}^{d\times n}$ and $\mathbb{R}^d$ such that

(*i*) $(Y, Z) \in \mathcal{M}_p^c(\beta, a, T)$, that is

$$\mathbb{E}\left(\sup_{0\leq t\leq T} e^{p\beta A(t)}|Y_t|^{2p} + \left(\int_0^T e^{\beta A(t)} a^2(t)|Y_t|^2 dt\right)^p + \left(\int_0^T e^{\beta A(t)}|Z_t|^2 dt\right)^p\right) < +\infty.$$

(*ii*) $\{Y_t : t \in [0, T]\}$ is a continuous process with values in $\overline{Dom(\Phi)}$.

(*iii*) $\{K_t : t \in [0, T]\}$ is a continuous process with $K_0 = 0$ a.s. and absolutely continuous with respect to the measure $e^{\beta A(t)} a^{-2}(t)dt$.

(iv) For every $t \in [0,T]$,

$$Y_t = \xi + \int_t^T f(s, Y_s, Z_s)\, ds - \int_t^T Z_s dW_s + K_T - K_t\ .$$

(v) For every optional couple of processes (α, β) such that $(\alpha_t, \beta_t) \in Gr\ (\partial\Phi)$,

$$\int_0^T \langle Y_s - \alpha_s, dK_s + a^2(s)\beta_s ds\rangle \leq 0\ .$$

In the sequel, we may and do assume without loss of generality that $\Phi(x) \geq \Phi(0) = 0$. So, we have $(0,0) \in Gr\ (\partial\Phi)$.
Now, let us recall some facts about Yosida approximation of sub-differential operators. For every $n \in \mathbb{N}^*$, and every $x \in \mathbb{R}^d$, let

$$\Phi_n(x) = \min_{y \in \mathbb{R}^d} \left\{ \frac{n}{2} |x - y|^2 + \Phi(y) \right\}.$$

Let $J_n(x)$ be the unique solution of the differential inclusion $x \in J_n(x) + \frac{1}{n}\partial\Phi(J_n(x))$ (see Barbu, Precupanu [3]).
Let us note that Φ_n and J_n satisfy the following:

$i)$ Φ_n: $\mathbb{R}^d \longrightarrow \mathbb{R}$ is a convex function of class C^1 with Lipschitz derivative.

$ii)$ For every $x \in \mathbb{R}^d$, $\nabla\Phi_n(x) = n(x - J_n(x)) := A_n(x)$.

$iii)$ There exist $a \in interior(Dom(\Phi))$ and positive numbers R, C such that for every $z \in \mathbb{R}^d$ and every $n \in \mathbb{N}^*$

$$\langle \nabla\Phi_n(z)^*, z - a\rangle \geq R \mid A_n(z) \mid - C \mid z \mid - C \text{ for all } n \in \mathbb{N}^*.$$

$iv)$ For every $x \in \mathbb{R}^d$ $A_n(x) \in \partial\Phi(J_n(x))$.

The map J_n is called the resolvent of the monotone operator $\partial\Phi$ and A_n is the Yosida approximation of $\partial\Phi$. A_n is a Lipschitz continuous and monotone function.
We also have

$$\text{for every } x \in \mathbb{R}^d, \quad \inf_{y \in \mathbb{R}^d} \Phi(y) \leq \Phi(J_n x) \leq \Phi_n(x) \leq \Phi(x). \tag{4.1}$$

We recall a sub-differential stochastic inequality due to Pardoux and Rascanu [15].

Proposition 4.2. *(Pardoux and Rascanu [15].) Let $\Psi : \mathbb{R}_+ \times \mathbb{R}^d \to \mathbb{R}$ be a function of class C^1 such that for every $t \geq 0$, $\Psi(t,.) : \mathbb{R}^d \to \mathbb{R}$ is a convex function. Let $\{X_t; t \geq 0\}$ be a continuous semi-martingale. Then, $\forall s \leq t$:*

$$\Psi(s, X_s) + \int_s^t \left[\frac{\partial\Psi(r, X_r)}{\partial r} dr + \langle \nabla_x \Psi(r, X_r), dX_r\rangle \right] \leq \Psi(t, X_t),\ \mathbb{P} - p.s.$$

Now, for every $n \in \mathbb{N}^*$, let us consider the non reflected BSDE:

$$Y_t^n = \xi + \int_t^T [f(s, Y_s^n, Z_s^n) - a^2(s) A_n(Y_s^n)] ds - \int_t^T Z_s^n dW_s\ ,\ \forall t \in [0, T\,]. \tag{4.2}$$

Since A_n is a monotone operator, the function $f_n(s,y,z) = f(s,y,z) - a^2(s)A_n(y)$ satisfies **(H1)**.
By virtue of Theorem 3.4, the BSDE (4.2) has an unique solution $(Y^n, Z^n) \in \mathcal{M}_p{}^c(\beta, a, T)$.
Now, we put

$$K_t^n = -\int_0^t a^2(s)A_n(Y_s^n)ds.$$

The main result of this section is

Theorem 4.3. *Assume* **(H1)-(H4)**. *Then the RBSDE* (T, ξ, f, Φ) *has a unique solution* (Y, Z, K) *in the space* $\mathcal{M}_p^c(\beta, a, T) \times L^2(\Omega)$. *Moreover*

$$\lim_{n \longrightarrow +\infty} \mathbb{E}\left(\sup_{0 \le t \le T} |K_t^n - K_t|^2 \right) = 0.$$

The proof of this result need the following lemmas.

Lemma 4.4. *Assume* **(H1)-(H4)**. *Then*

(i)

$$\sup_{n \in \mathbb{N}^*} \mathbb{E}\left\{ \left(\int_0^T e^{\beta A(t)} a^2(t) |Y_t^n|^2 dt \right)^p + \left(\int_0^T e^{\beta A(t)} |Z_t^n|^2 dt \right)^p dt \right\} < +\infty,$$

(ii)

$$\sup_{n \in \mathbb{N}^*} \mathbb{E}\left(\sup_{0 \le t \le T} e^{p\beta A(t)} |Y_t^n|^{2p} \right) < +\infty,$$

(iii)

$$\sup_{n \in \mathbb{N}^*} \mathbb{E}\left(\int_0^T e^{\beta A(t)} a^2(t) |A_n(Y_t^n)|^2 dt \right) < +\infty.$$

Proof. By virtue of Itô's formula, we have

$$\begin{aligned} & e^{\beta A(t)} |Y_t^n|^2 + \beta \int_t^T e^{\beta A(s)} a^2(s) |Y_s^n|^2 ds + \int_t^T e^{\beta A(s)} |Z_s^n|^2 ds \\ \le\ & e^{\beta A(T)} |\xi|^2 + 2\int_t^T e^{\beta A(s)} \langle Y_s^n, f(s, Y_s^n, Z_s^n) - a^2(s)A_n(Y_s^n) \rangle ds - 2\int_t^T e^{\beta A(s)} \langle Y_s^n, Z_s^n dW_s \rangle. \end{aligned} \tag{4.3}$$

Since $-\langle Y_s^n, a^2(s)A_n(Y_s^n)\rangle \le 0$ and

$$\begin{aligned} 2\langle Y_s^n, f(s, Y_s^n, Z_s^n)\rangle &\le 2\theta(s)|Y_s^n|^2 + |Y_s^n|[\nu(s)|Z_s^n| + |f(s,0,0)|] \\ &\le (2 + \frac{\beta}{2})a^2(s)\, |Y_s^n|^2 + \frac{1}{2}\, |Z_s^n|^2 + \frac{2}{\beta} \frac{|f(s,0,0)|^2}{a^2(s)}, \end{aligned}$$

By virtue of (4.3) and Proposition 3.2, we have

$$\begin{aligned} & \mathbb{E}\left[\left(\int_0^T e^{\beta A(t)} a^2(t) |Y_t^n|^2 dt \right)^p + \left(\int_0^T e^{\beta A(t)} |Z_t^n|^2 dt \right)^p \right] \\ \le\ & C\mathbb{E}\left[e^{p\beta A(T)}(1 + |\xi|^{2p}) + \frac{2^p}{\beta^p} \left(\int_0^T e^{\beta A(s)} \frac{|f(s,0,0)|^2}{a^2(s)} ds \right)^p \right]. \end{aligned}$$

Therefore (i) and (ii) are proved.
Let us deal with the proof of (iii). We put $\Psi_n(t, Y_t^n) = e^{\beta A(t)}\Phi_n(Y_t^n)$.
By virtue of Proposition 4.2, we have

$$e^{\beta A(t)}\Phi_n(Y_t^n) + \beta\int_t^T e^{\beta A(s)}a^2(s)\Phi_n(Y_s^n)ds + \int_t^T e^{\beta A(s)}\nabla\Phi_n(Y_s^n)dY_s^n \leq e^{\beta A(T)}\Phi_n(Y_T^n).$$

Since $\Phi_n \geq 0$, we deduce that

$$\begin{aligned}
& e^{\beta A(t)}\Phi_n(Y_t^n) + \int_t^T e^{\beta A(s)}a^2(s)|A_n(Y_s^n)|^2ds \\
\leq\ & e^{\beta A(T)}\Phi_n(Y_T^n) + \int_t^T e^{\beta A(s)}|A_n(Y_s^n)||f(s,Y_s^n,Z_s^n)|ds - \int_t^T e^{\beta A(s)}\langle A_n(Y_s^n), Z_s^n dW_s\rangle. \\
\leq\ & e^{\beta A(T)}\Phi_n(Y_T^n) + \frac{1}{2}\int_t^T e^{\beta A(s)}a^2(s)\,|A_n(Y_s^n)|^2ds + \frac{1}{2}\int_t^T e^{\beta A(s)}\frac{|f(s,Y_s^n,Z_s^n)|^2}{a^2(s)}ds \\
-\ & \int_t^T e^{\beta A(s)}\langle A_n(Y_s^n), Z_s^n dW_s\rangle.
\end{aligned}$$

It follows that

$$\begin{aligned}
& \mathbb{E}\left(e^{\beta A(t)}\Phi_n(Y_t^n) + \frac{1}{2}\int_t^T e^{\beta A(s)}a^2(s)|A_n(Y_s^n)|^2ds\right) \\
\leq\ & \mathbb{E}\left(e^{\beta A(T)}\Phi_n(\xi)\right) + \frac{1}{2}\mathbb{E}\left(\int_t^T e^{\beta A(s)}\frac{|f(s,Y_s^n,Z_s^n)|^2}{a^2(s)}ds\right).
\end{aligned}$$

Now in view of (i),(ii), **(H1)**, **(H3)** and **(H4)**, we obtain that

$$\sup_{0\leq t\leq T}\mathbb{E}\left(\Phi_n(Y_t^n)e^{\beta A(t)}\right) \leq C,$$

which leads to (iii). □

Proposition 4.5. *Assume* **(H1)-(H4)**. *Then,* (Y^n, Z^n) *is a Cauchy sequence in* $\mathcal{M}^c(\beta, a, T)$.

Proof. By virtue of Itô's formula , we have :

$$\begin{aligned}
& e^{\beta A(t)}\,|Y_t^n - Y_t^m|^2 + \int_t^T e^{\beta A(s)}\,|Z_s^n - Z_s^m|^2ds + \beta\int_t^T e^{\beta A(s)}a^2(s)\,|Y_s^n - Y_s^m|^2ds \\
=\ & 2\int_t^T e^{\beta A(s)}\langle Y_s^n - Y_s^m, f(s,Y_s^n,Z_s^n) - f(s,Y_s^m,Z_s^m)\rangle ds - 2\int_t^T e^{\beta A(s)}\langle Y_s^n - Y_s^m, (Z_s^n - Z_s^m)dW_s\rangle \\
& -2\int_t^T e^{\beta A(s)}\langle Y_s^n - Y_s^m,\, A_n(Y_s^n) - A_m(Y_s^m)\rangle ds.
\end{aligned}$$

Since,

$$I = J_n + \frac{1}{n}A_n = J_m + \frac{1}{m};A_n(Y_s^n) \in \partial\Phi(J_n(Y_s^n));A_m(Y_s^m) \in \partial\Phi(J_m(Y_s^m)) \text{ and } ab \leq \frac{1}{4}a^2 + b^2,$$

one can show that

$$-\langle Y_s^n - Y_s^m,\, A_n(Y_s^n) - A_m(Y_s^m)\rangle \leq \frac{1}{4m}|A_n(Y_s^n)|^2 + \frac{1}{4n}|A_m(Y_s^m)|^2.$$

So, in view of Burkhölder-Davis-Gundy's inequality and similar arguments as in the proof of Lemma 4.4-(ii), one can prove that for β large enough

$$\mathbb{E}\left(\sup_{0\le t\le T} e^{\beta A(t)}\,|Y_t^n-Y_t^m|^2\right)+\mathbb{E}\left(\int_0^T e^{\beta A(s)}\,|Z_s^n-Z_s^m|^2ds\right)$$
$$+\mathbb{E}\left(\int_0^T e^{\beta A(s)}a^2(s)\,|Y_s^n-Y_s^m|^2ds\right)\le C\left(\frac{1}{n}+\frac{1}{m}\right).$$

Therefore $\{(Y^n,Z^n):n\in\mathbb{N}^*\}$ is a Cauchy sequence in $\mathcal{M}^c(\beta,a,T)$. □

The following Lemma is due to Saisho [18]

Lemma 4.6. *Let* $(k^n:n\ge 1)$ *and* $(y^n:n\ge 1)$ *be two sequences in* $C([0,T],\mathbb{R}^d)$ *which converge uniformly to* k *and* y *respectively. Assume that* k^n *is of bounded variation and such that* $\sup_{n\in N^*}\|k^n\|_T<+\infty$, *where* $\|.\|_T$ *stands for the total variation on* $[0,T]$. *Then*

$$\int_0^T \langle y_s^n, dk_s^n\rangle \longrightarrow \int_0^T \langle y_s, dk_s\rangle.$$

Proof of the Theorem 4.3.
Existence. Proposition 4.5 implies that the sequence $\{(Y^n,Z^n):n\in\mathbb{N}^*\}$ has a limit (Y,Z) in the space $\mathcal{M}^c(\beta,a,T)$.
Now, Fatou's lemma and Lemma 4.4 ensure that $(Y,Z)\in\mathcal{M}_p^c(\beta,a,T)$.
Let us put

$$\begin{aligned}K_t^n &= Y_0^n-Y_t^n-\int_0^t f(s,Y_s^n,Z_s^n)ds+\int_0^t Z_s^n dW_s\\ K_t &= Y_0-Y_t-\int_0^t f(s,Y_s,Z_s)ds+\int_0^t Z_s dW_s.\end{aligned}$$

One can show that

$$\lim_{n\to+\infty}\mathbb{E}\left(\sup_{0\le t\le T}|K_t^n-K_t|^2\right)=0.$$

Therefore the sequence $\{(K_n):n\in\mathbb{N}^*\}$ converges uniformly in $L^2(\Omega)$ to K as $n\to+\infty$. Now, we shall prove that the triple (Y,Z,K) is the solution of the RBSDE (T,ξ,f,Φ). Taking a subsequence if necessary, we may assume that

$$\sup_{0\le t\le T}|K_t^n-K_t|\longrightarrow 0 \text{ as } n\to+\infty.$$

It follows that K is a continuous process.
Moreover,in view of Lemma 4.4, we deduce that

$$\sup_{n\in N^*}\mathbb{E}\|K^n\|_{H^1(0,T,\,dL_s\,\mathbb{R}^d)}<+\infty,$$

where $dL_s=e^{\beta A(s)}a^{-2}(s)ds$ and $H^1(0,T,dL_s,\mathbb{R}^d)$ is the Sobolev space of all absolutely continuous functions with derivative in $L^2([0,T]\,,dL_s)$.
Hence, the sequence $\{(K_n):n\in\mathbb{N}^*\}$ is bounded in the Hilbert space $L^2(\Omega,H^1(0,T,dL_s,\mathbb{R}^d))$.

Therefore, there exists a subsequence of the sequence $\{(K_n) : n \in \mathbb{N}^*\}$ which converges weakly as $n \longrightarrow +\infty$. The limit K belongs to $L^2(\Omega, H^1(0,T,dL_s,\mathbb{R}^d))$ and for almost all $\omega \in \Omega$, $K_.(\omega) \in H^1(0,T,dL_s,\mathbb{R}^d)$.
It follows that K is absolutely continuous and

$$-dK_t = V_t dL_t \text{ with } V_t \in \partial\Phi(Y_t).$$

The continuity of the process Y comes from the convergence of (Y^n) in the space $L^{2,c}(\beta, a, [0,T], \mathbb{R}^d)$.

Now, we show that

$$\mathbb{P}\{Y_t \in \overline{Dom(\Phi)}\} = 1, \quad \forall \leq t \leq T.$$

In view of the convergence of the sequence $\{(Y^n)\}$ in the space $L^{2,c}(\beta, a, [0,T], \mathbb{R}^d)$, there exists a subsequence $\{Y^{n_k} : k \in \mathbb{N}^*\}$ such that $\forall\, t \in [0,T]$, $Y_t^{n_k} \longrightarrow Y_t$ a.s. as $k \longrightarrow +\infty$
Therefore, $Y_t \in \overline{Dom(\Phi)}$ for every $t \in [0,T]$.
We end the proof of the existence result by showing that (v) of Definition 4.1 holds. To this end, we use Lemma 4.6.
Indeed, we have

$$\sup_{n\in N^*} \mathbb{E}\left(\int_0^T e^{\beta A(s)}\, a^2(s)|A_n(Y_s^n)|^2 ds\right) < +\infty.$$

Taking a subsequence, we may assume that $\{K^n : n \in \mathbb{N}^*\}$ and $\{Y^n : n \in \mathbb{N}^*\}$ converge uniformly to K and Y respectively as $n \longrightarrow +\infty$. For every optional process (α, β) such that $(\alpha_s, \beta_s) \in Gr(\partial\Phi)$, we have

$$\int_0^T \langle J_n(Y_s^n) - \alpha_s, dK_s^n + a^2(s)\beta_s ds\rangle = \int_0^T a^2(s)\langle J_n(Y_s^n) - \alpha_s, -A_n(Y_s^n) + \beta(s)ds\rangle \leq 0.\text{a.s}\ .$$

By passing to the limit and using Lemma 4.6, we obtain (v).

Uniqueness. Let (Y,Z,K) and (Y',Z',K') be two solutions. Since $\partial\Phi$ is a monotone operator, $-\frac{dK_t}{dL_t} \in \partial\Phi(Y_t)$ and $-\frac{dK_t'}{dL_t} \in \partial\Phi(Y_t')$, one can verify that

$$\mathbb{E}\left(\int_t^T e^{\beta A(s)}\, \langle Y_s - Y_s', dK_s - dK_s'\rangle\right) \leq 0.$$

Therefore, by applying Itô's formula to $e^{\beta A(t)}|Y_t - Y_t'|^2$ and using Young's inequality, one can prove that if (Y,Z,K) (resp.(Y',Z',K')) is solution of the RBSDE$((T,\xi,f,\Phi))$ (resp.(T,ξ',f',Φ)) then

$$\begin{aligned}\mathbb{E}\,(e^{\beta A(t)}|Y_t - Y_t'|^2) + \frac{1}{2}\mathbb{E}\left(\int_t^T e^{\beta A(s)}\, |Z_s - Z_s'|^2 ds\right) + C\mathbb{E}\left(\int_t^T e^{\beta A(s)} a^2(s)\, |Y_s - Y_s'|^2 ds\right)\\ \leq \mathbb{E}\,(e^{\beta A(t)}|\xi - \xi'|^2) + \frac{2}{\beta}\mathbb{E}\left(\int_t^T e^{\beta A(s)} \frac{|f(s,Y_s',Z_s') - f'(s,Y_s',Z_s')|^2}{a^2(s)} ds\right),\end{aligned}$$

which leads to the uniqueness of the solution.

References

[1] K. Bahlali, A. Elouaflin, M. N'ZI, (2004), Backward stochastic differential equations with stochastic monotone coefficients. *J. Appl. Math. Stoch. Anal.* 2004: 4, 317-335.

[2] K. Bahlali, E. Essaky, Y. Ouknine, (2004), Reflected backward stochastic differential equations with locally monotone coefficient, *Stoch. Anal. Appl.*, **22**, 4, 939–970.

[3] V. Barbu, T. Precupanu, Convexity and Optimization in Banach Spaces (Mathematics and Its Applications. East European Series) 1986.

[4] Ph. Briand, R. Carmona, (2000), BSDEs with polynomial growth generators. *J. Appl. Math. Stoch. Anal.* **13**, no. 3, 207-238.

[5] R. Darling, E. Pardoux, (1997), Backward SDE with random terminal time and application to semilinear PDE. *Annals of Probab*. **25** (4), 1135-1159.

[6] N. El Karoui, and S.J Huang, (1997), A general result of existence and uniqueness of backward stochastic differential equations, in *Backward Stochastic Differential Equations. Editors N. El Karoui and L. Mazliak. Pitman Research Notes in Mathematics Series* vol.364, 27-36. *Longman.*

[7] N. El Karoui, S. Peng, M. Quenez, (1997), Backward stochastic differential equations in finance. *Finance* **7**(1) , 1-71 .

[8] N.El Karoui, C.Kapoudjian, E.Pardoux, S.Peng, M.C.Quenez, (1997), Reflected solutions of backward SDE's and related obstacle problems for PDE's. *Annals Probab.* **25***(2)*, 702-737.

[9] S. Hamadène, J.P. Lepeltier, (1995), Backward equations, stochastic control and zero-sum stochastic differential games. *Stoch. Stoch. Reports* **54** , 221-231.

[10] M. Kobylanski, (2000), Backward stochastic differential equations and partial differential equations with quadratic growth. *Ann. Probab.*, **28**, no. 2, 558–602.

[11] M. N'zi ,Y.Ouknine, (1997), Equations différentielles stochastiques rétrogrades multivoques. *Probability and Mathematical Statistics* vol.17 , Fasc.2 , pp 259-275.

[12] E. Pardoux, S. Peng, (1990), Adapted solution of a backward stochastic differential equation. *Syst Cont Lett.* **14** , 55-61.

[13] S. Peng, (1991), Probabilistic interpretation for systems of quasilinear parabolic partial differential equations. *Stochastics Stochastics Rep.*, **37**, no. 1-2, 61–74.

[14] E. Pardoux, S. Peng, (1992), Backwrad SDE and quasilinear parabolic PDEs, in " *Stochastic partial differential equations and their applications " , B.L. Rosovskii and R. Sowers eds LNCIS* **176**, 200-217 . Springer, New York.

[15] E. Pardoux and A. Rascanu, (1998), Backward stochastic differential equations with subdifferential operator and related variational inequalities. *Stoch. Proc. Appl.* **76**, 191-215.

[16] E. Pardoux, (1998), Backward stochastic differential equations and viscosity solutions of systems of semilinear parabolic and elliptic PDEs of second order in L. Decreuseufond, J. Gjerde, B. Oksendal and A.S Ustunel, Editors, *Stochastic analysis and related topics. The Geilo worksop, 1996. Birkhaússer.*

[17] E. Pardoux, (1999), BSDEs, weak convergence and homogenization of semilinear PDEs in *F. H Clarke and R. J. Stern (eds.), Nonlinear Analysis, Differential Equations and Control,* 503-549 . Kluwer Academic Publishers.

[18] Y. Saisho, (1987), Stochastic differential equations for multidimensional domains with reflecting boundary. *Probab. Theory Rel. Fields* **74**, 455-477.

In: African Diaspora Mathematics Research Progress
Editor: Toka Diagana, pp. 75–81
ISBN 978-1-60456-204-0
© 2008 Nova Science Publishers, Inc.

Chapter 5

SUBMANIFOLDS OF THE UNIT SPHERE

Bazanfaré Mahaman*
Département de Mathématiques et Informatique
Université Abdou Moumouni, B.P. 10662, Niamey, Niger

Abstract

In this paper we establish a pinching condition to insure that submanifolds of codimension $p \geq 2$ in the unit sphere are spheres.

AMS Subject Classification: 53C40, 53C42.

Keywords: Parallel mean curvature vector field, second fundamental form , unit sphere and pinching constant.

1 Introduction

Let N be an oriented submanifold of codimension p in the unit sphere S^{n+p} with parallel mean curvature vector field. Denote by H the mean curvature and by S the square length of the second fundamental form. One of the fundamental questions is to have a positive constant called a pinching constant on the length of second fundamental form, depending only on n and the codimension of the submanifold N. In [13] and [12] S.T.Yau and H. W . Xu respectively established a pinching constant depending only on n and p. G. Chen et X. Zou [2] studied the case $p > 2$ and showed that if $2 \leq n \leq 7$ and $S \leq (2/3)n$ then N is totally umbilic, hence is a sphere . Unfortunately the constants are not the best ones. For $p = 1$ in [4] and for $p = 2$ in [5], Z. H. Hou gived a best possible constant under the condition that S is constant or N is compact without boundary. The same results are simultaneously obtained by M.J. Wang and S.J. Li in [11]

The purpose of this paper is the following theorem:

Theorem 1.1 *Let N^n be an n-dimensional connected submanifold in the unit sphere S^{n+p} with parallel mean curvature vector field. If*

$$a) \quad \sup_{x \in N} S < C(n,p) \tag{1}$$

*E-mail: bmahaman@yahoo.fr

where

$$C(n,p) = \frac{2n}{1 - \frac{1}{p-1} + \sqrt{\frac{(n-2)^2}{n-1} + \left(\frac{3p-4}{p-1}\right)^2}}$$

$$C(n,2) = 2\sqrt{n-1}$$

then N^n is a sphere with radius $r = \sqrt{n/(n+S)}$.

b) For $p \geq 2$ if N is compact and

$$S \leq C(n,p) \tag{2}$$

then N is the totally umbilic sphere $S^n\left(\sqrt{n/(n+S)}\right)$ or an hypersphere of the totally geodesic S^{n+1} in S^{n+p}.

For minimal submanifolds, S. S. Chern, M. P. Do Carmo et S. Kobayashi [1] obtained the constant $\frac{n}{2-\frac{1}{p}}$ and classified such submanifolds.

2 Preliminaries

We suppose that $p \geq 2$.(For $p = 1$ see for example [4]). Let x be a point in N and (e_i, ξ_α) $1 \leq i \leq n$, $1 \leq \alpha \leq p$ be a local orthonormal frame field on the sphere S^{n+p} such that (e_i) lies in the tangent bundle TN and (ξ_α) lies in the normal bundle $\nu(N)$ of N. For each α, $1 \leq \alpha \leq p$, define a linear map

$A_\alpha : T_xN \rightarrow T_xN$ by:

$$\langle A_\alpha X, Y \rangle = \langle \nabla_X Y, e_\alpha \rangle.$$

Thus we deduce a map A_α such that, for all vector fields X and Y on N, we have in a neighborhood of x,

$$\langle A_\alpha X, Y \rangle = \langle \nabla_X Y, e_\alpha \rangle.$$

The mean curvature vector h is defined by

$$h = \frac{1}{n} \sum_{\alpha=1}^{p} (traceA_\alpha)\xi_\alpha$$

and the square length of the second fundamental form is

$$S = \sum_{\alpha=1}^{p} trace(A_\alpha^2). \tag{3}$$

Set $H = \|h\|$ and $\xi_1 = \frac{h}{H}$. Let φ_α defined by:

$$\langle \varphi_\alpha X, Y \rangle = \langle X, Y \rangle \langle h, \xi_\alpha \rangle - \langle A_\alpha X, Y \rangle.$$

Hence

$$\varphi_1 = HId - A_1; \quad \varphi_\alpha = -A_\alpha \quad \forall \alpha \geq 2. \tag{4}$$

Let

$$\varphi \ : \ T_xM \times T_xM \longrightarrow T_xM$$
$$(X,Y) \longmapsto \sum_{\alpha=1}^{p} \langle \varphi_\alpha(X), Y\rangle \xi_\alpha. \qquad (5)$$

3 Proof

In [7] Okumura proved:

$$\frac{1}{2}\Delta S = \sum_{\alpha=1}^{p} |\nabla A_\alpha|^2 + nS - \sum_{\alpha=1}^{p} (traceA_\alpha)^2 + \sum_{\alpha,\beta\geq 2} trace\left([A_\alpha, A_\beta]\right)^2$$
$$+ \sum_{\alpha,\beta=1}^{p} (traceA_\alpha)\left(traceA_\alpha A_\beta^2\right) - \sum_{\alpha,\beta=1}^{p} \left(traceA_\alpha A_\beta\right)^2. \qquad (6)$$

Since ξ_1 is parallel, we deduce, by the choice of the φ_α the formula:

$$\frac{1}{2}\Delta|\varphi|^2 = \sum_{\alpha=1}^{p} |\nabla\varphi_\alpha|^2 + n(1+H^2)|\varphi|^2 - nH\sum_{\alpha=1}^{p}(trace\varphi_1\varphi_\alpha^2)$$
$$- \sum_{\alpha,\beta=1} \left(trace\varphi_\alpha\varphi_\beta\right)^2 + \sum_{\alpha,\beta\geq 2} trace\left([\varphi_\alpha, \varphi_\beta]\right)^2. \qquad (7)$$

The following lemma was proved by W. Santos [9]:

Lemma 3.1 *Let $A, B : R^n \rightarrow R^n$ be two symmetric linear maps such that $[A,B] = 0$ and $traceA = traceB = 0$ then:*

$$-\frac{n-2}{\sqrt{n(n-1)}}(traceA^2)(traceB^2)^{1/2} \leq traceA^2B \leq \frac{n-2}{\sqrt{n(n-1)}}(traceA^2)(traceB^2)^{1/2} \quad (8)$$

and the equality holds on the right hand (resp. left hand) side if and only if $n-1$ of the eigenvalues x_i of A and the corresponding eigenvalues y_i of B satisfy:

$$|x_i| = \frac{(traceA^2)^{1/2}}{\sqrt{n(n-1)}}, \quad x_ix_j \geq 0 \qquad (9)$$

and

$$y_i = \frac{(traceB^2)^{1/2}}{\sqrt{n(n-1)}}(resp.y_i = -\frac{(traceB^2)^{1/2}}{\sqrt{n(n-1)}}). \qquad (10)$$

Then, she deduce for $p \geq 2$:

$$\frac{1}{2}\Delta|\varphi|^2 \geq |\varphi|^2\left(n + nH^2 - \frac{n(n-2)}{\sqrt{n(n-1)}}|\varphi||h| - \frac{2p-3}{p-1}|\varphi|^2\right)$$
$$+ \sum_{\alpha=1}^{p} |\nabla\varphi_\alpha|^2 + |\varphi_1|^2\left(2|\varphi|^2 - |\varphi_1|^2\right). \qquad (11)$$

Thus from (11), we get:

$$\frac{1}{2}\Delta|\varphi|^2 \geq |\varphi|^2\left(n+nH^2-\frac{n(n-2)}{\sqrt{n(n-1)}}H|\varphi|-\frac{2p-3}{p-1}|\varphi|^2\right) + \sum_{\alpha=1}^{p}|\nabla\varphi_\alpha|^2+|\varphi_1|^2\left(2|\varphi|^2-|\varphi_1|^2\right). \quad (12)$$

For $a>0$ we have:

$$\sqrt{n}H|\varphi| \leq \frac{1}{2}\left(anH^2+\frac{1}{a}|\varphi|^2\right). \quad (13)$$

Hence

$$\frac{1}{2}\Delta|\varphi|^2 \geq |\varphi|^2\left(n+nH^2-\frac{n(n-2)}{2\sqrt{(n-1)}}aH^2-\frac{(n-2)}{2a\sqrt{n-1}}|\varphi|^2 - \frac{2p-3}{p-1}|\varphi|^2\right) \quad (14)$$

$$\geq |\varphi|^2\left[n+nH^2\left(1-\frac{n-2}{2\sqrt{n-1}a}+\frac{n-2}{2a\sqrt{n-1}}+\frac{2p-3}{p-1}\right) - \left(\frac{n-2}{2a\sqrt{n-1}}+\frac{2p-3}{p-1}\right)S\right]. \quad (15)$$

$$\frac{1}{2}\Delta|\varphi|^2 \geq |\varphi|^2\left[n+\frac{nH^2}{a}\left(-\frac{n-2}{2\sqrt{n-1}}a^2+\frac{3p-4}{p-1}a+\frac{n-2}{2\sqrt{n-1}}\right) - \left(\frac{n-2}{2a\sqrt{n-1}}+\frac{2p-3}{p-1}\right)S\right]. \quad (16)$$

Take $a>0$ such that:

$$-\frac{n-2}{2\sqrt{n-1}}a^2+\frac{3p-4}{p-1}a+\frac{n-2}{2\sqrt{n-1}}=0 \quad (17)$$

Hence

$$a=e^{\theta} \quad \text{where } \theta = Argsh\frac{(3p-4)\sqrt{n-1}}{(p-1)(n-2)}. \quad (18)$$

It follows that

$$\frac{1}{2}\Delta|\varphi|^2 \geq |\varphi|^2\left\{n-\left(\frac{(n-2)e^{-\theta}}{2\sqrt{n-1}}+\frac{2p-3}{p-1}\right)S\right\}. \quad (19)$$

By elementaries computations we take

$$e^{-\theta} = \sqrt{\frac{(3p-4)^2(n-1)}{(p-1)^2(n-2)^2}+1}-\frac{3p-4)\sqrt{n-1}}{(p-1)(n-2)}$$

Hence if

$$S \leq C(n,p) = \frac{2n}{1-\frac{1}{p-1}+\sqrt{\frac{(n-2)^2}{n-1}+\left(\frac{3p-4}{p-1}\right)^2}}$$

then

$$\Delta|\varphi|^2 \geq 0.$$

The following lemma is due to Omori [8]

Lemma 3.2 *Let N be a n dimensional complete Riemannian manifold with Ricci curvature bounded from below. If* $h : N \longrightarrow \mathbb{R}$ *is a* C^∞ *function such that* $\sup_N h < +\infty$, *then there exists a suite* $(x_k) \subset N$ *so that:*

$$i) \quad \lim_k h(x_k) = \sup h \tag{20}$$

$$ii) \quad \lim_k dh(x_k) = 0 \tag{21}$$

$$iii) \quad \lim_k \Delta h(x_k) \leq 0 \tag{22}$$

If N is as above, by Gauss-Codazzi equation we have:

$$\begin{aligned} Ric(e_i) &= (n-1) + \sum_{\alpha=1}^{n} (traceA_\alpha)\langle A_\alpha e_i, e_i\rangle - \sum_\alpha |A_\alpha(e_i)|^2 \\ &\geq (n-1) - n^2H^2 - S \end{aligned} \tag{23}$$

for any orthonormal vectors field e_i on N.

Proof of the theorem 1.1 a
Suppose that $\sup_N S < C(n,p)$; then there exists an $\varepsilon > 0$ such that

$$S \leq C(n,p) - \varepsilon.$$

Hence

$$\frac{1}{2}\Delta|\varphi|^2 \geq |\varphi|^2(n - \frac{n(C(n,p)-\varepsilon)}{C(n,p)} \geq \frac{n\varepsilon}{C(n,p)}|\varphi^2| \geq 0. \tag{24}$$

If $|\varphi| \neq 0$ then there exists $u_0 \in M$ so that $|\varphi|^2(u_0) = b > 0$. Hence $\sup_N |\varphi|^2 \geq |\varphi|^2(u_0) > b$ and by (24) we have $\frac{1}{2}lim_k\Delta|\varphi|^2(x_k) \geq \frac{n\varepsilon}{C(n,p)}b > 0$ which contradicts iii) lemma 3.2. One concludes that $|\varphi|^2 = 0$ which means that N is the totally umbilic sphere $S^n(\sqrt{\frac{n}{n+S}})$.

Proof of the theorem1.1 b)
Suppose $S \leq C(n,p)$ and N is compact. Hence by the maximum principe we $\Delta|\varphi|^2 = 0$ and $|\varphi|^2$ is constant. In addition we have:

$$|\varphi|^2\left(n - \frac{n}{C(n,p)}S\right) + |\varphi_1|^2\left(2|\varphi|^2 - |\varphi_1|^2\right) + \sum_\alpha |\nabla\varphi_\alpha|^2 \geq 0. \tag{25}$$

Hence either $|\varphi|^2 = 0$ and consequently N is a sphere or else

$$S = C(n,p) \qquad \text{and} \qquad |\varphi_1|^2 = 0 \tag{26}$$

which implies that $A_1 = HId$ and

$$\varphi(X,Y) = \sum_{\alpha=2}^{p} \langle \varphi_\alpha(X), Y\rangle \xi_\alpha.$$

Hence $\varphi_h = 0$.

If $S = C(n,p)$ then the inequalities (11) and (12) become equalities, and we deduce that $|\varphi| = 0$ and the conclusion follows.

In [12] W. Xu established the following proposition:

Proposition 3.3 *Let N be a p-codimensional, compact submanifold of the sphere S^{n+p} with parallel mean curvature vector field and $p > 1$. If*

$$S < \min\left\{\frac{2n}{1+\sqrt{n}}, \frac{n}{2-1/(p-1)}\right\}$$

and if the Gauss function of N is relatively affine, then N is a standard hypersphere in the totally geodesic sphere S^{n+1} as a submanifold of S^{n+p}.

If $p \leq \frac{2\sqrt{n}+6}{6+1/(n-1)} + 1$ then by elementaries computations one can show that $C(n,p) > \frac{2n}{1+\sqrt{n}}$.

References

[1] S. S. Chern, M. P. Do Carmo and S. Kobayashi, Minimal submanifolds of a sphere with second fundamental form of constant length,In *Functional Analysis and Related Fields* (edited by F. Brower), Springer-Verlag, Berlin, 1970, 59-65.

[2] G. Chen and X. Zou, *Rigidity of compact manifolds in a unit sphere* , *Kodai Math. J.* **18** (1995) 75-85.

[3] T. Hasanis, Characterisation of totally umbilical hypersurfaces, *Proceeding of Amer. Math. Society volume* **81** n° 3 1981.

[4] Z. H. Hou, Hypersurfaces in a sphere with constant mean curvature, *Proc. Amer. Math. Soc.* **125**(1997) pp. 1193-1196.

[5] Z. H. Hou, A pinching problem on submanifolds with parallel mean curvature vector field in a sphere, *Kodai MATH. J.* **21** (1998)35-45.

[6] K. Nomizu and B. Smyth, A formula of Simon's type and hypersurfaces with constant mean curvature , *J Differential Geometry*, **3**,1969 p.367-377.

[7] M. Okumura, Hypersurfaces and a pinching problem on the Sobolev fundamental tensor, *Amer. J. Math* **96** ,1974 p. 207-213.

[8] H. Omori, Isometric immersions of Riemannian manifolds, *J. Math. Soc. Japan* **19** (1967) pp.205–214.

[9] W. Santos, Submanifolds with parallel mean curvature vector in spheres, *Tôhoku Math. J.* **46**(1994),403-415.

[10] J. Simon, Minimal varieties in Riemannian manifolds, *Ann. of Math.* **88** , 1968 p.62-105.

[11] M. J. Wang and S. J. Li, Submanifolds with parallel mean curvature vector in a sphere, *Kodai Math. J.* **21** (1998), 201-207

[12] H. W. Xu, A pinching constant of Simon's type and isometric immersion, *Chinese Ann.of Math. Ser.A* **12**(1991) pp.261–269.

[13] S. T. Yau, Submanifolds with constant mean curvature II, *Amer. J. Math.* **97** (1975) pp.76–100.

In: African Diaspora Mathematics Research Progress
Editor: Toka Diagana, pp. 83–90
ISBN 978-1-60456-204-0
© 2008 Nova Science Publishers, Inc.

Chapter 6

COLLAPSE OF A VOID SPHERICAL BUBBLE IMMERSED IN A NON-NEWTONIAN FLUID

Célestin Wafo Soh*
Department of Mathematics College of Science Engineering
and Technology Jackson State University JSU Box 17619
1400 J R Lynch St., Jackson, MS 39217, USA

Abstract

We study analytically the dynamics of a single void spherical bubble immersed in a power-law non-Newtonian fluid and in a second-grade fluid. We derive the equation of motion of the bubble wall and we prove that it is integrable. We establish that near collapse, the radius of the bubble behaves like $(t_c - t)^k$, where $k \in \{2/5, (2-n)/2\}$ for a power-law fluid of index n, $k \in \{1/2, 1/3\}$ for a second-grade fluid, and t_c is the collapse time.

AMS Subject Classification: 76B10, 76A05, 34-xx.

Keywords: cavitation, collapse time, power-law fluid, second-grade fluid.

1 Introduction

In several industrial processes, bubbles or cavities occur by design or accident. They are most feared when they appear accidentally, are damaging, and cannot be controlled. For instance, the performance and safety of pulsed lasers used in surgery for ablation are affected by bubbles formation and collapse. Indeed, the drilling and cutting of soft tissues in liquid surroundings are followed by the formation of fast-expanding and imploding cavities that may cause side effects such as arterial dilation in angioplasty [12], and injury to the retina during vitreoretinal membrane removal [8]. In several devices such as pumps and propellers, the formation and collapse of bubbles generates a lot of noises and causes damage to components. Particularly, the noise generated by cavitation compromises the ability of submarines to hide from enemies.

*E-mail: wafosoh@yahoo.com, celestin.wafo@jsums.edu

Although cavitation has damaging effects in some circumstances, it is a boon in some applications. It is often use in food processing industries to homogenize, mix and break down suspended particles in colloidal liquids such as milk. Water purification plants use cavitation to break down pollutants and organic molecules. Perhaps one of the most spectacular use of cavitation is in sonoluminescence [6, 9, 3] where short bursts of light emanate from imploding bubbles in a liquid when excited by sound.

In most cases, bubbles form in a region of a fluid where the absolute pressure is less than or equal the vapor pressure of the fluid. Several methods are available to generate cavities. These methods include pressure variation by flow over submerged bodies, pressure variation by vibrating boundaries, pressure variation in vortex motion of a fluid, direct variation of the hydrostatic pressure and temperature variation by heating a liquid.

Arguably, the first consistent statement of the problem of the collapse of a single bubble surrounded by a fluid appeared in Besant's book [1]: "An infinite mass of homogeneous incompressible fluid acted upon by no force is at rest, and a spherical portion of the fluid is suddenly annihilated; it is required to find the instanteneous alteration of pressure at any point of the mass, and the time in which the cavity will fill up, the pressure at an infinite distance being supposed to remain constant". An analytical solution to this problem in the case of inviscid fluid was proposed by Lord Raleigh [10]. Raleigh's solution provides a basis for the explanation of damages to solid boundaries by collapsing cavitation bubbles.

In this paper, we shall focus on the collapse of a spherical bubble immersed in a non-Newtonian fluid due to initial bubble wall motion. A similar problem was recently tackled in the case of Newtonian fluids by Bogoyavlenskiy [2]. The outline of this paper is the following. Section 2 is concerned with the derivation of equations of motion of a spherical bubble immersed in a power-law non-Newtonian fluid and a second-grade fluid. In Section 3, we study analytically the equations obtained in the previous section. This study allows us to provide collapse criteria, collapse time and behavior near collapse.

2 Dynamics of a Void Spherical Bubble Surrounded by a Non-Newtonian Fluid

We consider a void spherical bubble submerged in an infinite non-Newtonian fluid. The bubble is assumed to remain spherical during deformation. Owing to the symmetry of the problem, we adopt spherical coordinates (r, ϕ, θ) with origin at the center of the bubble. The fluid motion is governed by Navier-Stokes equations. The latter have the following non trivial components.

$$\frac{\partial T_{rr}}{\partial r} + \frac{2T_{rr} - T_{\theta\theta} - T_{\phi\phi}}{r} = \rho\left(\frac{\partial u}{\partial t} + \frac{1}{2}\frac{\partial u^2}{\partial r}\right) \tag{2.1}$$

$$\frac{\partial u}{\partial r} + 2\,\frac{u}{r} = 0, \tag{2.2}$$

where u is the radial fluid velocity, T is the stress tensor of the liquid and ρ its density which is assumed constant.

Following Bogoyavlenskiy [2], we assume that the ambient pressure and the surface tension on the bubble wall are negligible. Thus

$$T_{rr}|_{r=R(t)} = 0, \quad T_{rr}|_{r=\infty} = 0, \tag{2.3}$$

where $R(t)$ is the radius of the deforming bubble at time t.

We further assume that initially the bubble has a given velocity:

$$R|_{t=0} = R_0, \quad \left.\frac{dR}{dt}\right|_{t=0} = -V_0, \tag{2.4}$$

where $R_0 > 0$ and $V_0 > 0$ are constants. In order to solve the problem (2.1)-(2.5), we need a constitutive relation i.e. we need to specify the stress tensor $\underline{T}$. We consider in the sequel two important cases: power-law non-Newtonian and second-grade fluids.

2.1 Case of Power-Law Fluids

The constitutive relation for power-law fluids is

$$\underline{T} = -p\underline{I} + 2\mu \left|\frac{1}{2}\mathrm{tr}(\underline{D}^2)\right|^{\frac{n-1}{2}} \underline{D}, \tag{2.5}$$

where p is the mechanical pressure, $\mu > 0$ and $n > 0$ are constants, $\underline{I}$ is the identity matrix, and $\underline{D}$ is the rate of deformation tensor. In our case, the latter is given by

$$\underline{D} = \mathrm{diag}\left(\frac{\partial u}{\partial r}, \frac{u}{r}, \frac{u}{r}\right), \tag{2.6}$$

and the non-zero component of the stress tensor are

$$T_{rr} = -p + 2\mu\left(\frac{u^2}{r^2} + \frac{1}{2}\left(\frac{\partial u}{\partial r}\right)^2\right)^{\frac{n-1}{2}} \frac{\partial u}{\partial r} \tag{2.7}$$

$$T_{\theta\theta} = T_{\phi\phi} = -p + 2\mu\left(\frac{u^2}{r^2} + \frac{1}{2}\left(\frac{\partial u}{\partial r}\right)^2\right)^{\frac{n-1}{2}} \frac{u}{r} \tag{2.8}$$

Integrate the incompressibility condition (2.2) and use the fact that $u|_{r=R(t)} = \dot{R}(t)$ to obtain

$$u = \dot{R}\left(\frac{R}{r}\right)^2, \tag{2.9}$$

where the dot stands for differentiation with respect to t.

Substitute Eqs. (2.7)-(2.8) and Eq. (2.9) into (2.1), integrate both sides of the resulting equation with respect to r from $R(t)$ to ∞ and use the boundary conditions (2.3) to obtain

$$R\ddot{R} + \frac{3}{2}\dot{R}^2 + \frac{3^{\frac{n-1}{2}}4\mu}{n\rho}\left(\frac{\dot{R}}{R}\right)^n = 0. \tag{2.10}$$

2.2 Case of Second-Grade Fluids

The stress tensor for second-grade fluids is [11]

$$\underline{T} = -p\underline{I} + \mu\underline{A_1} + \alpha_1\underline{A_2} + \alpha_2\underline{A_1}^2, \tag{2.11}$$

where p is the mechanical pressure, μ is the viscosity, and α_1 and α_2 are α_2 are normal stress moduli,

$$\underline{A_1} = \text{grad } \underline{v} + (\text{grad } \underline{v})^T, \tag{2.12}$$

$$\underline{A_2} = \frac{d\underline{A_1}}{dt} + \underline{A_1} \text{ grad } \underline{v} + (\text{grad } \underline{v})^T \underline{A_1}, \tag{2.13}$$

where d/dt stands for the total time derivative and $\underline{v}$ the velocity field. It has been argued by Dunn and Fosdick [4] that in order to assure thermodynamics compatibility the additional conditions

$$\mu \geq 0, \quad \alpha_1 \geq 0, \quad \alpha_1 + \alpha_2 = 0 \tag{2.14}$$

have to impose. For a recent discussion of restrictions (2.14), the reader is referred to the review by Dunn and Rajagopal [5].

For the flow under consideration, after some calculations, we find that the nontrivial component of the stress tensor are:

$$T_{rr} = -p - 4\mu\frac{\dot{R}R^2}{r^3} + 16(\alpha_1+\alpha_2)\frac{\dot{R}^2R^4}{r^6} - 4\alpha_1\frac{\ddot{R}R^2 + 2\dot{R}^2R}{r^3}, \tag{2.15}$$

$$T_{\theta\theta} = T_{\phi\phi} = -p + 2\mu\frac{\dot{R}R^2}{r^3} + 4(\alpha_1+\alpha_2)\frac{\dot{R}^2R^4}{r^6} + 2\alpha_1\frac{\ddot{R}R^2 + 2\dot{R}^2R}{r^3}. \tag{2.16}$$

Substitute Eqs. (2.15)-(2.16) and Eq. (2.9) into (2.1), integrate both sides of the resulting equation with respect to r from $R(t)$ to ∞ and use the boundary conditions (2.3) to obtain

$$\left(R + \frac{\alpha_1}{\rho R}\right)\ddot{R} + \left(\frac{3}{2} + \frac{\alpha_1 - \alpha_2}{\rho R}\right)\dot{R}^2 + 4\mu\frac{\dot{R}}{R} = 0. \tag{2.17}$$

3 Solution and Analysis of the Equations of Motion

Our aim in this section is to solve the Eqs. (2.11) and (2.17) subject to the initial conditions (2.5). When possible, we shall determine the collapse criteria and time.

3.1 Motion of a Spherical Bubble in Power-Law Non-Newtonian Fluid

Here we are concerned with solving Eq. (2.11) subject to the initial condition (2.5). we start by non-dimensionalizing Eq. (2.11). For this purpose, we introduce the dimensionless variables

$$\tilde{R}=\frac{R}{R_0},\quad \tilde{t}=t\,\frac{V_0}{R_0},\quad \tilde{\mu}=\frac{\mu}{\frac{3^{\frac{n-1}{2}}V_0^{n-2}}{n\rho R_0^n}}. \tag{3.1}$$

In the new variables (3.1), Eq. (2.11) reads

$$\tilde{R}\tilde{R}''+\frac{3}{2}\tilde{R}'^2+4\tilde{\mu}\left(\frac{\tilde{R}'}{R}\right)^n=0, \tag{3.2}$$

where the prime stands for differentiation with respect to $\tilde{t}$. The initial conditions (2.5) become

$$\tilde{R}(0)=1,\quad \tilde{R}'(0)=-1. \tag{3.3}$$

In order to solve Eq. (3.2), we introduce the change of variables

$$x=\tilde{R},\quad y=\frac{d\tilde{t}}{d\tilde{R}}\,. \tag{3.4}$$

Under the change of variables (3.4), Eq. (3.2) becomes

$$y'-\frac{3}{2x}y=\frac{4\tilde{\mu}}{x^{n+1}}\,y^{3-n}, \tag{3.5}$$

where the prime stands for differentiation with respect to x. We have to solve Eq.(3.5) subject to the condition

$$y(1)=-1. \tag{3.6}$$

We consider the following cases.

Case 1: $n=2$. In this case Eq. (3.5) is linear and the solution solution to the problem (3.5)-(3.6) is

$$y=-x^{3/2}e^{2\tilde{\mu}(1-x^{-2})}. \tag{3.7}$$

Thus

$$\tilde{t}=\int_{\tilde{R}}^{1}x^{3/2}e^{2\tilde{\mu}(1-x^{-2})}\,dx\,. \tag{3.8}$$

The dimensionless collapse time is given by

$$\tilde{t}_c=\int_0^1 x^{3/2}e^{2\tilde{\mu}(1-x^{-2})}\,dx=\frac{e^{2\tilde{\mu}}}{2}\,\Gamma(-5/4,2\tilde{\mu}). \tag{3.9}$$

where Γ is the incomplete Gamma function [7].

Case 2: $n\neq 2$. In this case , Eq. (3.5) is a Bernoulli equation that can be solved by introducing the new dependent variable

$$z=y^{n-2}. \tag{3.10}$$

In terms of z, Eq.(3.5) is the linear first-order ordinary differential equation

$$z'-\frac{3(n-2)}{2x}z=\frac{4\tilde{\mu}(n-2)}{x^{n+1}}. \tag{3.11}$$

Using an appropriate integrating factor, we may rewrite Eq. (3.11) as

$$\frac{d}{dx}\left[x^{3(2-n)/2}\, z\right] = 4\tilde{\mu}(n-2)x^{(4-5n)/2}. \tag{3.12}$$

The integration of Eq. (3.12) leads to the following subcases.

Subcase 2.1: $n = 6/5$. Solving Eq. (3.12) and using Eqs. (3.6) and (3.10), we find that

$$\tilde{t} = \int_{\tilde{R}}^{1} x^{3/2}\left(1 - \frac{16\tilde{\mu}}{5}\ln x\right)^{-5/4} dx. \tag{3.13}$$

The dimensionless collapse time is

$$\tilde{t}_c = \int_{0}^{1} x^{3/2}\left(1 - \frac{16\tilde{\mu}}{5}\ln x\right)^{-5/4} dx = \frac{1}{4}\left(\frac{5}{4\tilde{\mu}}\right)^{5/4} e^{\frac{5}{4\tilde{\mu}}}\Gamma\left(-1/4, \frac{5}{4\tilde{\mu}}\right). \tag{3.14}$$

Subcase 2.2: $n \neq 6/5, 2$ **.** In this case, after solving Eq.(3.12) and imposing the (3.6), we obtain

$$\tilde{t} = \left(\frac{8\tilde{\mu}(n-2)}{6-5n}\right)^{\frac{1}{n-2}} \int_{\tilde{R}}^{1} x^{\frac{n}{2-n}}\left(\alpha x^{\frac{5n-6}{2}} + 1\right)^{\frac{1}{n-2}} dx \tag{3.15}$$

where

$$\alpha = \frac{(6-5n)(-1)^{n-2}}{8\tilde{\mu}(n-2)} - 1. \tag{3.16}$$

Note that in the Newtonian case i.e. when $n = 1$, the integral appearing in Eq. (3.15) can be expressed in terms of elementary functions and the result of Bogoyavlenskiy [2] is obtained.

The spherical bubble will collapse provided

$$0 < \tilde{t}_c = \left(\frac{8\tilde{\mu}(n-2)}{6-5n}\right)^{\frac{1}{n-2}} \int_{0}^{1} x^{\frac{n}{2-n}}\left(\alpha x^{\frac{5n-6}{2}} + 1\right)^{\frac{1}{n-2}} dx < \infty \tag{3.17}$$

As $\tilde{R} \to 0$, the dynamics of the bubble is described by

$$R = \begin{cases} \left(\frac{5}{2\alpha^{\frac{1}{n-2}}}\right)^{2/5}\left(\frac{8\tilde{\mu}(n-2)}{6-5n}\right)^{2/5}(\tilde{t}_c - \tilde{t})^{2/5} & \text{if } n < 6/5 \\ \left(\frac{2}{2-n}\right)^{\frac{2-n}{2}}\left(\frac{6-5n}{8\tilde{\mu}(n-2)}\right)^{1/2}(\tilde{t}_c - \tilde{t})^{\frac{2-n}{2}} & \text{if } n > 6/5;\ n \neq 2. \end{cases} \tag{3.18}$$

3.2 Motion of a Spherical Bubble Second-Grade Non-Newtonian Fluid

Introduce the change of variable $x = R$ and $y = dt/dR$. In terms of the new variables, Eq. (2.17) reads

$$y' - \frac{3\rho x^2 + 2(\alpha_1 - \alpha_2)}{2x(\rho x^2 + \alpha_1)}y - \frac{4\mu}{\rho x^2 + \alpha_1}y^2 = 0. \tag{3.19}$$

Equation (3.19) is a Bernoulli equation. In order to solve it we introduce the change of dependent variable $z = 1/y$ and Eq. (3.19) is transformed into

$$\frac{d}{dx}\left(zx^{1-\alpha}(x^2+\beta)^{\frac{2\alpha+1}{4}}\right) = -4\mu\beta x^{1-\alpha}(x^2+\beta)^{\frac{2\alpha-3}{4}}, \tag{3.20}$$

where

$$\alpha = \frac{\alpha_2}{\alpha_1}, \quad \beta = \frac{\alpha_1}{\rho} \cdot \tag{3.21}$$

We choose $\alpha = -1$ to enforce thermodynamics compatibility (2.14) and exclude the Newtonian case $\alpha_1 = 0 = \alpha_2$. with this choice, Eq. (3.20) yields after integration and non-dimensionalization

$$\frac{d\tilde{R}}{d\tilde{t}} = -\delta(1+\gamma)^{-1/4}\left(\tilde{R}^2+\gamma\right)^{1/4}\tilde{R}^{-2} + (\delta-1)\tilde{R}^{-1} - (\delta-1)\left(\tilde{R}^2+\gamma\right)^{1/4}\tilde{R}^{-2}\int_1^{\tilde{R}}(x^2+\gamma)^{-1/4}dx \tag{3.22}$$

where

$$\delta = 1 + \frac{8\mu}{\beta V_0 R_0}, \quad \gamma = \frac{\beta}{R_0^2} \cdot \tag{3.23}$$

Equation (3.22) is a separable equation. However it is not possible to write its general solution in terms of elementary functions.

When $\delta = 1$ i.e. $\mu = 0$,

$$\tilde{t} = (1+\gamma)^{1/4}\int_{\tilde{R}}^1 x^2(x^2+\gamma)^{-1/4}\, dx, \tag{3.24}$$

and the non-dimensional collapse time is

$$\tilde{t}_c = (1+\gamma)^{1/4}\int_{\tilde{0}}^1 x^2(x^2+\gamma)^{-1/4}\, dx = \frac{1}{2}\left(1+\frac{1}{\gamma}\right)^{1/4} F\left(\frac{1}{4},\frac{3}{2};\frac{5}{2},-\frac{1}{\gamma}\right). \tag{3.25}$$

From (3.25) we infer that

$$\lim_{\gamma\to\infty} \tilde{t}_c = \frac{1}{2} \cdot \tag{3.26}$$

Now we study the bubble motion as $\tilde{R}$ approaches zero in the case where $\delta \neq 1$. From Eq.(3.22) we infer that as $\tilde{R} \to 0$,

$$R = \begin{cases} \left[3\delta - 3(\delta-1)(1+\gamma)^{1/4}K\right]^{1/3}(\tilde{t}_c-\tilde{t})^{1/3} & \text{if } \delta-(\delta-1)(1+\gamma)^{1/4}K \neq 0 \\ \sqrt{2(\delta-1)}(\tilde{t}_c-\tilde{t})^{1/2} & \text{otherwise,} \end{cases} \tag{3.27}$$

where $K = \int_0^1 (x^2+\gamma)^{-1/4}$.

4 Conclusion

We have studied analytically the dynamics due to an initial wall motion of a void spherical bubble immersed into a power-law non-Newtonian fluid, and a second-grade fluid. We systematically showed that the equation of motion is integrable. Further we obtained explicitly the dynamics near collapse i.e. when the radius of the bubble approaches zero. We establish that near collapse, the radius of the bubble behave like $(t_c - t)^k$, where $k \in \{2/5, (2-n)/2\}$ for a power-law fluid of index n, $k \in \{1/2, 1/3\}$ for a second-grade fluid , and t_c is the collapse time. This result is important since near collapse, numerical schemes are hampered by dire numerical instabilities: knowing the behavior near collapse guide in strategies to mitigate instabilities.

References

[1] W. H. Besant, *A treatise on hydrostatics and hydrodynamics* , Deighton Bell, Cambridge, 1859.

[2] V. A. Bogoyavlenskiy, Differential criterion of a bubble collapse in viscous liquids, *Phys. Rev. E,* **60** (1999), pp. 504-508.

[3] M. Brenner, S. Hilgenfeldt, and D. Lohse, Single bubble sonoluminescence, *Rev. Mod. Phys.*,**74** (2002), pp. 425-484.

[4] J. E. Dunn and R. L. Fordick, Thermodynamics, stability, and boundedness of fluids of complexity 2 and fluids of second grade, *Arch. Rational Mech. Anal.*, **56** (1974), pp. 191-252.

[5] J. E. Dunn and K. R. Rajagopal, Fluids of differential type: critical review and thermodynamic analysis, *Intl. J. Engng. Science,* **33** (1995), pp. 689-729.

[6] D. F. Gaitan, L. A. Crum, R. A. Roy, and C. C. Church, *J. Acoust. Soc. Am.* **91** (1992), pp. 3166-3183.

[7] I. S. Gradshteyn, I. M. Ryzhik and A. Jeffrey, *Table of integrals, series, and product*, Academic Press, Boston, 1994.

[8] T. I. Margolis, D. A. Farmath, M. Destro, and C. A. Puliafito, Erbium-YAG laser surgery on experimental vitreous membranes, *Arch. Ophtalmol.,* **107** (1989), pp. 424-428.

[9] S. J. Putterman, Sonoluminescence: sound into light, *Scientific American,* **272** (1995), pp. 46-51.

[10] L. Rayleigh, On the pressure developed in a liquid during the collapse of a spherical cavity, *Phil. Mag.*, **34** (1917), pp. 94-98.

[11] C. Trussdell and W. Noll, The non-linear field theory of mechnaics, *Hanbuch der Physik,* vol. 3, Springer, Berlin, 1965.

[12] T. G.Van Leeuwen, J. H. Meertens, E. Velema,and M. J. Post, Intraluminal vapor bubble induced by excimer laser pulse causes microsecond arterial dilation and invagination leading to extensive wall damage in the rabbit, *Circulation*, **87** (1993), pp. 1258-1263.

In: African Diaspora Mathematics Research Progress
Editor: Toka Diagana, pp. 91–104

ISBN 978-1-60456-204-0
© 2008 Nova Science Publishers, Inc.

Chapter 7

HAZARD RATE PREDICTION IN LIFE TIME DATA ANALYSIS

*Kossi Essona Gneyou**
Department of Mathematics University of Lome Lome, BP 1515, TOGO

Abstract

We consider in this paper a nonparametric estimation of the hazard rate function based on right-censored data using the wavelets method. Asymptotic properties and strong uniform consistency rates are established under suitable conditions.

AMS Subject Classification: 62G05; 62G20.

Keywords: Hazard rate, life time data, right censorship model, wavelets method.

1 Introduction

Let $X_1, X_2, \ldots, X_n$ be a sequence of independent, identically distributed (i.i.d) non-negative random variables (r.v.) with common continuous distribution function (d.f.) F and density function f and $Y_1, Y_2, \ldots, Y_n$ another sequence of i.i.d non-negative r.v. with common continuous d.f. G, both sequences (X_i) and (Y_i) being defined on the same probability space $(\Omega, \mathcal{A}, \mathbb{P})$ and mutually independent. In this paper we are concerned with the nonparametric estimation of the hazard rate function λ defined by

$$\lambda(t) = \frac{f(t)}{1 - F(t)}, \qquad F(t) < 1 \tag{1.1}$$

whenever F and f are unknown and the observations available are the pairs (Z_i, δ_i) where for $i = 1, 2, \ldots, n$

$$Z_i = \min(X_i, Y_i), \qquad \delta_i = \mathbf{1}_{\{X_i \le Y_i\}} = \begin{cases} 1 & \text{if } X_i \le Y_i \\ 0 & \text{otherwise.} \end{cases} \tag{1.2}$$

*E-mail: kgneyou@tg.refer.org

X_i is said to be censored on the right by Y_i when $\delta_i = 0$.

Set $X = X_1$, $Y = Y_1$, $Z = Z_1$, $\delta = \delta_1$ and denote by H the d.f. of Z. It is easily seen that $H(t) = 1-(1-F(t))(1-G(t))$.

In life testing, medical follow-up and other studies, the random variables X and Y indicate respectively, the observation of the occurrence of an event of interest (as failure time or onset of AIDS time) and the occurrence of another event (called a censoring event). It is the case for example, in medical follow-up of persons infected with HIV (human immunodeficiency virus), where Y is the age of a patient and X is his onset age of AIDS, unknown except $X > Y$.

The hazard rate as the probability density or the probability distribution function, is a basic characteristic describing the behavior of a random variable X. The problem of nonparametric estimation of the hazard rate function is related to that of the density function. The most important methods considered in this topic by statisticians, in non-censored or censored models as well, are kernels, nearest neighbor, orthogonal series or projection methods. For progress and developments in the literature, see e.g. Watson and Leadbetter[27] and [28] , Deheuvels [5], Földes, Rejető and Winter [9], Stute [25], Tanner and Wong [26], Mielniczuk [23], Gneyou [10], [11], [12], and [13], Müler and Wang [24], Deheuvels and Einmhal [6], and the references therein.

The aim of this paper is to give another approach to estimate the hazard rate function λ based on wavelets method. Indeed, wavelets method and multiresolution analysis of $L^2(\mathbb{R})$ introduced by Mallat[21] (see also Meyer [22]), has become during these last years, a mathematical sound tool for adaptively estimating functions. A remarkable property of wavelet transform is to reflect the local regularity of the original function, being large where the function is irregular and small where the function is smooth. For references upon density and hazard rate estimation using wavelets method, see e.g. Kerkyacharian and Picard [20], Hall and Patil[15], and recently, Antoniadis, Grégoire and Nason [1], Aubin and Massiani [2]. Optimal rates of convergence of the mean integrated squared error, L^p-loss, and weak convergence, have been investigated by these researchers.

To get our wavelet estimator of λ, we use the projection approach. Recall that any function $f \in L^2(\mathbb{R})$ may be expressed in the form

$$f(t) = \sum_{k\in\mathbb{Z}} \alpha_{j_0k}\phi_{j_0k}(t) + \sum_{j\geq j_0}\sum_{k\in\mathbb{Z}} \beta_{jk}\psi_{jk}(t), \tag{1.3}$$

where $\{\phi_{j_0k}\}_{k\in\mathbb{Z}}$ and $\{\psi_{jk}\}_{k\in\mathbb{Z}}$ form an orthonormal basis of $L^2(\mathbb{R})$ and are defined by

$$\phi_{j_0k}(t) = 2^{j_0/2}\phi(2^{j_0}t-k), \qquad \psi_{jk}(t) = 2^{j/2}\psi(2^{j}t-k). \tag{1.4}$$

The coefficients α_{j_0k} and β_{jk} are given by

$$\alpha_{j_0k} = \int_{-\infty}^{+\infty} f(x)\phi_{j_0k}(x)dx \quad \text{and} \qquad \beta_{jk} = \int_{-\infty}^{+\infty} f(x)\psi_{jk}(x)dx, \tag{1.5}$$

where ϕ is the scaling function or father wavelet satisfying $\int_{-\infty}^{+\infty}\phi(t)dt = 1$ and ψ is a the associate mother wavelet.

The decomposition in (1.3) gives an approximation of $f \in L^2(\mathbb{R})$ at resolution j_0 and the detail in f at resolutions finer than j_0.

To estimate the hazard rate function by wavelet method, we first fix an a priori resolution j depending on the sample size n and then obtain, under appropriate assumptions on the regularity of the density and the scaling functions, a linear wavelet estimator of the hazard rate, by projecting onto the subspace V_j of the multiresolution analysis (see e.g. Kerkyacharian and Picard[19] in the framework of the probability density estimation). At the end, we choose the optimal resolution (the smoothing parameter) $j_0 = j_0(n)$ by the classical method of minimizing the mean integrated squared error.

2 Estimation of the Hazard Rate Function by Wavelet Method

Let $T_F = \sup\{t \in \mathbb{R}^+ / F(t) < 1\}$, T_G and T_H defined as T_F with F replaced by G and H respectively. Obviously $T_H = \min(T_F, T_G)$ and it can be proved that $Z_{(n)} = \max(Z_1, \ldots, Z_n)$ tends to T_H almost surely as n tends to $+\infty$ (see Carbonez et al.[3]).

Let ϕ be the scaling function of the multiresolution analysis $(V_j)_{j\in\mathbb{Z}}$ of $L^2([0, T_F])$ satisfying the following assumptions :

(A1) ϕ is bounded with support in $[-L, L]$ and $\int_0^{T_F} \phi(x)dx = 1$.

(A2) ϕ is **r-regular** ($r \in \mathbb{N}^*$) i.e., ϕ is of class C^r, and ϕ and all its derivatives up to the order r are rapidly decreasing.

(A3) The family $\{\phi_{jk}\}_{k\in\mathbb{Z}}$ defined in (1.4) is an orthonormal basis of V_j.

Let P_{V_j} be the orthogonal projector into V_j. Then from the assumption $(A3)$, the kernel of P_{V_j} can be written as $\hat{K}_j(x,y) = 2^j \hat{K}(2^j x, 2^j y)$, where

$$\hat{K}(x,y) = \sum_{k\in\mathbb{Z}} \phi(x-k)\phi(y-k). \tag{2.1}$$

Under assumptions $(A1)$ - $(A3)$, $\hat{K}$ satisfies the following statements (see e.g. in Härdle et al.[16]) which are determining for the asymptotic properties of the estimator :

(S1) $\hat{K}$ is bounded i.e. there exists a positive constant C such that $\left|\hat{K}(x,y)\right| \le C$ for all $x, y \in [0, T_F]$.

(S2) $\hat{K}(x,y) = 0$ for $|\, x-y \,| > 2L$ and $\int_0^{T_F} \hat{K}(x,y)dy = \sum_{k\in\mathbb{Z}} \phi(x-k) = 1$.

(S3) For all positive integer $l \le r$, $\int_0^{T_F} (x-y)^l \hat{K}(x,y)dy = 0$.

Then, as in [19] (see in the proof of the Theorem 3.3 below), we define at resolution $j = j(n)$, a nonparametric wavelet estimator λ_n of the hazard rate λ by

$$\lambda_n(t) = \frac{1}{n}\sum_{i=1}^{n} \hat{K}_j(t, Z_i)\frac{\delta_i}{1-\hat{H}_n(Z_i)}, \qquad t \le Z_{(n)}, \tag{2.2}$$

where $1-\hat{H}_n = (1-\hat{F}_n)(1-\hat{G}_n)$ and $\hat{F}_n$ and $\hat{G}_n$ are Kaplan-Meier[18] product-limit estimators of F and G respectively given by (see e.g. in Diehl and Stute[7])

$$1-\hat{F}_n(t) = \begin{cases} \displaystyle\prod_{\{i/Z_{(i)} \le x\}} \left(\frac{n-i}{n-i+1}\right)^{\delta_{(i)}}, & \text{if } t \le Z_{(n)} \\ 0 & \text{if } t > Z_{(n)} \end{cases} \tag{2.3}$$

$1-\hat{G}_n(t)$ defined as $1-\hat{F}_n(t)$ with $\delta_{(i)}$ replaced in the formula (2.3) by $1-\delta_{(i)}$, where $Z_{(1)} \leq Z_{(2)} \leq \ldots \leq Z_{(n)}$ are the order statistics of the sample $(Z_1, Z_2, \ldots, Z_n)$ and for $i = 1, \ldots, n$, $\delta_{(i)}$ is the δ_j corresponding to $Z_{(i)} = Z_j$, $1 \leq j \leq n$.

In uncensored case (i.e. $G \equiv 0$), $\hat{H}_n$ is replaced in the formula (2.2) by the usual empirical distribution function of the X_is : $F_n(t) = \frac{1}{n}\sum_{i=1}^{n} 1_{\{X_i \leq t\}}$. If the censoring variable is known, $1-\hat{H}_n$ is replaced by $(1-\hat{F}_n)(1-G)$. For later proofs, we set, for all $t \in [0, T_F]$

$$\hat{E}\lambda_n(t) = \frac{1}{n}\sum_{i=1}^{n} \hat{K}_j(t, Z_i)\frac{\delta_i}{1-H(Z_i)}. \tag{2.4}$$

Note that, $\hat{E}\lambda_n(t)$ is not the mathematical expectation of $\lambda_n(t)$ but rather, is a r.v. whose expectation is the projection of $\lambda(t)$ on the linear subspace V_j spanned by $\{\phi_{jk}(t) = 2^{j/2}\phi(2^j t - k), k \in \mathbb{Z}\}$ and in the uncensored case, $\hat{E}\lambda_n(t) = \mathbb{E}(\lambda_n(t))$ where, $\mathbb{E}$ denotes the usual expectation.

The definition (2.2) of the wavelet estimator of λ differs to that of Antoniadis et al.[1]. Indeed, they have separately, estimated by wavelets method, the subdensity $f^*(t) = f(t)(1-G(t))$ of those observations that are still to fail and the probability $1-H$ of observations remaining at risk and thus, form an estimator of the hazard rate function by dividing the subdensity estimator by a wavelet estimator of $1-H$. More precisely, to estimate the subdensity f^*, they divided the time axis into a dyadic number of small intervals (bins) of equal width, binned the observed data into bins, and used wavelet regression estimate on the binned data to get an estimate of the subdensity.

Remark. Recalling that $\phi_{jk}(t) = 2^{j/2}\phi(2^j t - k), k \in \mathbb{Z}$, we can write

$$\lambda_n(t) = \sum_{k \in \mathbb{Z}} \alpha_{jk}\phi_{jk}(t), \tag{2.5}$$

where

$$\alpha_{jk} = \frac{1}{n}\sum_{i=1}^{n} \phi_{jk}(Z_i)\frac{\delta_i}{1-\hat{H}_n(Z_i)}. \tag{2.6}$$

Thus our wavelet estimator of the hazard rate function λ is linear and can be easily computed. Compared to formula (1.3), the second term on the right-hand side of the equality (1.3) can be neglected if one imposes a high degree of regularity on the density function f or on the hazard rate function λ directly. Because of the compactness of the support of (ϕ), the sum in (2.5) is finite.

In the next section, we study the asymptotic properties of our wavelet estimator and establish strong uniform consistency which is an extension of results obtained by the previous investigations in the topic.

3 Asymptotic Properties of the Wavelet Estimator of the Hazard Rate Function

Note that, the condition $G(T_F) < 1$ implies $T_H = T_F \leq T_G$ and then $H(T_F) < 1$. So, we will let this condition holds throughout the section. That enables us to establish the asymptotic properties and strong uniform consistency of the wavelet estimator on the whole interval $[0,T_F]$. Recall that, by assumptions $(A1)$ - $(A3)$, the statements $(S1)$ - $(S3)$ hold. We have

Theorem 3.1. *Suppose that $G(T_F)<1$, the density function $f\in C^1[0,T_F]$, the scaling function ϕ of the multiresolution analysis $(V_j)_{j\in\mathbb{Z}}$ of $L^2([0,T_F])$ satisfies assumptions $(A1)$ - $(A3)$ and $j=j(n)$ is a non-decreasing sequence such that $j(n)\longrightarrow+\infty$ and $n2^{-j(n)}\longrightarrow+\infty$ as $n\longrightarrow+\infty$. Then for all $t\in[0,T_F]$*

$$\mathbb{E}\lambda_n(t)=\lambda(t)+O(2^{-j(n)})+O((\frac{\log\log n}{n})^{\frac{1}{2}}). \tag{3.1}$$

If in addition, G has a bounded derivative $g=G'$ on $[0,T_G]$, then for all $t\in[0,T_F]$

$$\mathrm{Var}\lambda_n(t)=\frac{\lambda^*(t)}{n2^{-j(n)}}+O(2^{-j(n)}) \tag{3.2}$$

where

$$\lambda^*(t)=\frac{\lambda(t)}{1-H(t)}. \tag{3.3}$$

Proof: The proof of this theorem and the next are given in appendix.

Thus the estimator λ_n is consistent, with bias depending on j. So, an appropriate choice of j could be checked by having a small mean integrated squared error of λ_n (MISE(λ_n)). This is possible if one imposes a high order of regularity on the density function f. We have

Theorem 3.2. *Suppose that $G(T_F)<1$, f has a derivative of order $(r+1)$ and $f^{(r+1)}$ is bounded on $[0,T_F]$, ϕ satisfies assumptions $(A1)$ - $(A3)$, $j=j(n)$ is such that $j(n)\longrightarrow+\infty$ and $n2^{-j(n)}\longrightarrow+\infty$ as $n\longrightarrow+\infty$. Then*

$$\mathrm{MISE}(\lambda_n)=O(2^{-(2r+2)j(n)})+O(\frac{1}{n2^{-j(n)}})+O((\frac{\log\log n}{n})^{\frac{1}{2}}). \tag{3.4}$$

By the Theorem 3.2, it is easy to check that, if the multiresolution analysis $(V_j)_{j\in\mathbb{Z}}$ is r-regular and if $f^{(r+1)}$ exists and is bounded on $[0,T_F]$, then an optimal choice of j is $j\geq\frac{\log_2 n}{2r+3}$ which yields $\mathrm{MISE}(\lambda_n)=O(n^{-\frac{2r+2}{2r+3}})$, where $\log_2 n=\frac{\log n}{\log 2}$.

In the next theorem, we establish the strong uniform consistency of the wavelet estimator. We have

Theorem 3.3. *Let $T\leq T_F$ such that $H(T)<1$. Suppose that ϕ satisfies assumptions $(A1)$ - $(A3)$ and $j=j(n)$ is such that $j(n)\longrightarrow+\infty$, and $\frac{n2^{-2j(n)}}{\log\log n}\longrightarrow+\infty$ as $n\longrightarrow+\infty$. Then for n big enough, we have*

$$\sup_{0\leq t\leq T}|\lambda_n(t)-\lambda(t)|=O\left(\sqrt{\frac{\log\log n}{n2^{-2j(n)}}}\right)+O(2^{-j})\quad\text{a.s.} \tag{3.5}$$

The demonstration of Theorem 3.3 utilizes the process $\sqrt{n}(\Lambda_n-\Lambda)$ where Λ (resp. Λ_n) is the cumulative (resp. the Kaplan-Meier empirical cumulative) hazard function. Hence, according to the Csörgő's[4] approximation of the process $\sqrt{n}(\Lambda_n-\Lambda)$ by a Wiener process, one can obtain, from the limit set in (1.30) of Gu and Lai[14], a law of the iterated logarithm for the wavelet estimator λ_n. Namely, set for all $t \in [0, T_F]$, $d(t) = \int_0^t \lambda^*(s)ds = \int_0^t \frac{dF(s)}{(1-F(s))(1-H(s))}$. We have,

Theorem 3.4. *If the assumptions of the Theorem 3.3 are satisfied and $d(T) < +\infty$, then, the wavelet estimator λ_n satisfies*

$$\limsup_{n\to+\infty} \left(\frac{n}{2\Omega_{j(n)} 2^{j(n)} \log\log n} \right)^{\frac{1}{2}} \sup_{0\le t\le T} | \lambda_n(t) - \lambda(t) | = 1 \quad \text{a.s.,} \tag{3.6}$$

where $\Omega_{j(n)} = d(T) \inf_{0\le t\le T} \int_0^T \hat{K}^2(2^j t, s)ds.$

4 Appendix: Proofs

Before we prove the results of the previous section, we first need to establish the following lemmas :

Lemma 4.1. *Let $\hat{E}\lambda_n$ be defined as in (2.4) and suppose that $G(T_F) < 1$. Then*

$$\sup_{0\le t\le T_F} | \lambda_n(t) - \hat{E}\lambda_n(t) | = O(\sqrt{\frac{\log\log n}{n}}) \qquad \text{a.s.} \tag{4.1}$$

Proof: We have

$$\begin{aligned} | \lambda_n(t) - \hat{E}\lambda_n(t) | &= | \frac{1}{n}\sum_{i=1}^n \hat{K}_j(t, Z_i)\delta_i \left(\frac{1}{1-\hat{H}_n(t)} - \frac{1}{1-H(t)} \right) | \\ &\le \frac{1}{(1-\hat{H}_n(T_F))(1-H(T_F))} | \frac{1}{n}\sum_{i=1}^n \hat{K}_j(t, Z_i)\delta_i | \sup_{0\le t\le T_F} | \hat{H}_n(t) - H(t) | . \end{aligned} \tag{4.2}$$

Recalling that $(1-\hat{H}_n(x)) = (1-\hat{F}_n(t))(1-\hat{G}_n(t))$, where $\hat{F}_n$ and $\hat{G}_n$ are the Kaplan-Meier product-limit estimators of F and G respectively, we can write

$$\hat{H}_n(t) - H(t) = (\hat{F}_n(t) - F(t)) + (\hat{G}_n(t) - G(t)) - (\hat{F}_n(t) - F(t))\hat{G}_n(t) - (\hat{G}_n(t) - G(t))F(t).$$

Thus,

$$\sup_{0\le t\le T_F} | \hat{H}_n(t) - H(t) | \le 2[\sup_{0\le t\le T_F} | \hat{F}_n(t) - F(t) | + \sup_{0\le t\le T_F} | \hat{G}_n(t) - G(t) |].$$

Hence, making use of the result of Földes and Rejtő[8], we readily get

$$\sup_{0\le t\le T_F} | \hat{H}_n(t) - H(t) | = O(\sqrt{\frac{\log\log n}{n}}) \qquad \text{a.s.} \tag{4.3}$$

The lemma will be proved if we prove that, the first two terms in the right-hand side of (4.2) are bounded. Put $d = 1 - H(T_F) = (1 - F(T_F)(1 - G(T_F)) > 0$. From (4.3), there exists Ω_0 with $\mathbb{P}(\Omega_0) = 1$ such that, if $\omega \in \Omega_0$, then, for all n greater or equal to some $N_0(\omega)$, $1 - \hat{H}_n(T_F, \omega) > \frac{d}{2}$. Moreover, by strong law of large numbers applied to the i.i.d. r.v. $\hat{K}_j(t, Z_i)\delta_i, \ \ i = 1, \cdots, n$

$$\frac{1}{n}\sum_{i=1}^{n} \hat{K}_j(t, Z_i)\delta_i \longrightarrow m \quad \text{a.s.}$$

as $n \longrightarrow +\infty$ with,

$$m = \mathbb{E}(\hat{K}_j(t, Z)\delta) = \int_0^{T_F} \hat{K}_j(t, x)(1 - G(x))dF(x).$$

Since by assumption (A2) ϕ is regular, it follows that

$$\text{there exists } C > 0 \quad \text{such that} \quad | \hat{K}(t, x) | \leq C(1 + | t - x |)^{-2}. \tag{4.4}$$

Thus, (4.4) implies

$$| m | = | \int_{-2^j t}^{2^j (T_F - t)} \hat{K}(2^j t, 2^j t + u)(1 - G(t + 2^{-j}u)f(t + 2^{-j}u)du |$$

$$\leq \frac{2C \parallel f \parallel_\infty}{1 + 2^j \text{dist}((t, [0, T_F]))^2},$$

where $\text{dist}(t, A)$ denotes the distance between t and a subset A. Since $t \in [0, T_F]$, the quantity in the right-hand side of the last inequality is then bounded (by $2\,C \parallel f \parallel_\infty$). Consequently, m is bounded. This implies that $\frac{1}{n}\sum_{i=1}^{n} \hat{K}_j(t, Z_i)\delta_i$ is almost surely bounded i.e. there exists a constant $M > 0$ such that

$$| \frac{1}{n}\sum_{i=1}^{n} \hat{K}_j(t, Z_i)\delta_i | \leq M \quad \text{a.s.}$$

Thus, from (4.2) we have, for n sufficiently large,

$$\sup_{0 \leq t \leq T_F} | \lambda_n(t) - \hat{E}\lambda_n(t) | \leq \frac{2}{d^2} M \sup_{0 \leq t \leq T_F} | \hat{H}_n(t) - H(t) | \text{ a.s.} \tag{4.5}$$

The lemma follows from (4.3) and (4.5).

Lemma 4.2. *Suppose that $G(T_F) < 1$, G has a derivative $g = G'$ bounded on $[0, T_G]$ and that $f \in C^1[0, T_F]$. Then for all $t \in [0, T_F]$ and for $p = 1, 2$*

$$2^{(1-p)j} \int_0^{T_F} \hat{K}_j^p(t, x) \frac{\lambda(x)}{(1 - H(x))^{p-1}} dx = \frac{\lambda(t)}{(1 - H(t))^{p-1}} + O(2^{-j}). \tag{4.6}$$

Proof: By the property of ϕ and the orthonormal property of $\{\phi_{jk}, k \in \mathbb{Z}\}$, observe that,

$$\int_0^{T_F} \hat{K}_j(t, x)dx = \sum_{k \in \mathbb{Z}} 2^j \phi(2^j t - k) \int_0^{T_F} \phi(2^j x - k)dx$$

$$= \int_0^{T_F} \phi(u)du \sum_{k \in \mathbb{Z}} \phi(2^j t - k) = 1 \tag{4.7}$$

and

$$\int_0^{T_F} \hat{K}_j^2(t,x)dx = \int_0^{T_F} 2^{2j}(\sum_{k\in\mathbb{Z}} \phi(2^j t-k)\phi(2^j x-k))^2 dx$$
$$= 2^{2j}\int_0^{T_F} \sum_{k\in\mathbb{Z}} \phi^2(2^j t-k)\phi^2(2^j x-k)dx$$
$$= 2^j \int_0^{T_F} \phi^2(u)du \sum_{k\in\mathbb{Z}} \phi^2(2^j t-k) = 2^j. \tag{4.8}$$

Set, for all continuous function L,

$$\bar{L}(x) = 1-L(x), \quad \|L\|_\infty = \sup_{0\le t\le T_F} | L(t) | < +\infty \quad \text{and} \quad d = 1-F(T_F) > 0.$$

Then we can write

$$\lambda(x)-\lambda(t) = \frac{1}{\bar{F}(x)}(f(x)-f(t)) + \frac{f(x)}{\bar{F}(x)\bar{F}(t)}(F(x)-F(t)). \tag{4.9}$$

Recalling that $\lambda^*(s) = \dfrac{\lambda(s)}{\bar{H}(s)}$, we have

$$\lambda^*(x)-\lambda^*(t) = \frac{1}{V(x)}(f(x)-f(t)) + \frac{f(x)}{V(x)V(t)}(V(x)-V(t)), \tag{4.10}$$

where $V(x) = \bar{H}(x)\bar{F}(x) = \bar{G}(x)\bar{F}^2(x)$. So for $p=1$ and in view of (4.9), we have

$$| \int_0^{T_F} \hat{K}_j(t,x)\lambda(x)dx - \lambda(t) | = | \int_0^{T_F} 2^j \hat{K}(2^j t, 2^j x)(\lambda(x)-\lambda(t))dx |$$
$$\le \int_{-2^j t}^{2^j(T_F-t)} \hat{K}(2^j t, 2^j t+s) | \lambda(t+2^{-j}s)-\lambda(t) | ds$$
$$\le C_0 \|A_j\|_\infty \int_{-2L}^{2L} \hat{K}(2^j t, 2^j t+s)ds \tag{4.11}$$

where $C_0 = \max(d^{-1}, \| f \|_\infty d^{-2})$ and

$$A_j(t) = \sup_{0\le h\le 2^{-j}T_F} | f(t+h)-f(t) | + \sup_{0\le h\le 2^{-j}T_F} | F(t+h)-F(t) | .$$

By assumption on f, $\|A_j\|_\infty$ behaves like $O(2^{-j})$ so that

$$| \int_0^{T_F} \hat{K}_j(t,x)\lambda(x)dx - \lambda(t) | = O(2^{-j})$$

and the lemma is proved for $p=1$.

Likewise, we have for $p=2$,

$$| 2^{-j}\int_0^{T_F} \hat{K}_j^2(t,x)\lambda^*(x)dx - \lambda^*(t) | \le C_1 \|B_j\|_\infty \int_{-2L}^{2L} \hat{K}^2(2^j t, 2^j t+s)ds, \tag{4.12}$$

where $C_1 = \max(d^{-2}, \| f \|_\infty d^{-4})$ and

$$B_j(t) = \sup_{0 \le h \le 2^{-j} T_F} | f(t+h) - f(t) | + \sup_{0 \le h \le 2^{-j} T_F} | V(t+h) - V(t) | .$$

Moreover, by assumptions on f and g, it holds that, the derivative V' of V exists and is bounded on $[0, T_F]$. Thus $B(j)$ behaves like $O(2^{-j})$ and hence,

$$| 2^{-j} \int_0^{T_F} \hat{K}_j^2(t,x)\lambda^*(x)dx - \lambda^*(t) | = O(2^{-j}),$$

which is the statement of the lemma for $p = 2$.

Proof of Theorem 3.1: By Lemma 4.1, it is readily seen that, for all $t \in [0, T_F]$,

$$\mathbb{E}(\lambda_n(t)) = \mathbb{E}(\hat{E}\lambda_n(t)) + O(\sqrt{\frac{\log\log n}{n}}) \quad \text{and} \quad \mathrm{Var}\,\lambda_n(t) = \mathrm{Var}\,\hat{E}\lambda_n(t).$$

Thus, we have to evaluate the expectation and the variance of $\hat{E}\lambda_n(t)$. By Lemma 4.2

$$\begin{aligned} \mathbb{E}(\hat{E}\lambda_n(t)) &= \mathbb{E}\left[\hat{K}_j(t,Z)\frac{\delta}{1-H(Z)}\right] \\ &= \int_0^{T_F} \hat{K}_j(t,x)\frac{1-G(x)}{1-H(x)}dF(x) \\ &= \int_0^{T_F} \hat{K}_j(t,x)\lambda(x)dx = \lambda(t) + O(2^{-j}). \end{aligned} \tag{4.13}$$

Hence, $\mathbb{E}\lambda_n(t) = \lambda(t) + O(2^{-j}) + O(\sqrt{\frac{\log\log n}{n}})$ and (3.1) is proved.

For the variance term, we have

$$\begin{aligned} n\,\mathrm{Var}(\hat{E}\lambda_n(t)) &= \mathrm{Var}\left[\hat{K}_j(t,Z)\frac{\delta}{1-H(Z)}\right] \\ &= \mathbb{E}\left[\hat{K}_j^2(t,Z)\frac{\delta}{(1-H(Z))^2}\right] - \left(\mathbb{E}\left[\hat{K}_j(t,Z)\frac{\delta}{1-H(Z)}\right]\right)^2 \\ &= \mathrm{I} - \mathrm{II}. \end{aligned} \tag{4.14}$$

To conclude, it is enough to evaluate the two terms I and II. For the first term, we have

$$\begin{aligned} \mathrm{I} &= \mathbb{E}\left[\hat{K}_j^2(t,Z)\frac{\delta}{(1-H(Z))^2}\right] \\ &= \int_0^{T_F} \hat{K}_j^2(t,x)\frac{1-G(x)}{(1-H(x))^2}dF(x) \\ &= \int_0^{T_F} \hat{K}_j^2(t,x)\frac{\lambda(x)}{1-H(x)}dx \\ &= \int_0^{T_F} \hat{K}_j^2(t,x)\lambda^*(x)dx. \end{aligned} \tag{4.15}$$

Hence, by Lemma 4.2, it follows that $2^{-j}\mathrm{I} = \lambda^*(t) + O(2^{-j})$. As for the second term, we have

$$\mathrm{II} = \left(\mathbb{E}\left[\hat{K}_j(t,Z)\frac{\delta}{1-H(Z)}\right]\right)^2 = \left(\int_0^{T_F} \hat{K}_j(t,x)\lambda(x)dx\right)^2.$$

and the Lemma 4.2 implies that

$$\mathrm{II} = (\lambda(t) + O(2^{-j}))^2 = \lambda^2(t) + O(2^{-j}) + O(2^{-2j}) = \lambda^2(t) + O(2^{-j}).$$

It follows that,

$$n2^{-j}\,\mathrm{Var}(\hat{E}\lambda_n(t)) = \lambda^*(t) + O(2^{-j}),$$

which implies (3.2).

Proof of Theorem 3.2: By definition,

$$\mathrm{MISE}(\lambda_n) = \int_0^{T_F} \mathbb{E}(\lambda_n(t) - \lambda(t))^2 dt.$$

It follows that

$$\begin{aligned}\mathrm{MISE}(\lambda_n) &= \int_0^{T_F} \mathrm{Var}(\lambda_n(t))dt + \int_0^{T_F} (\mathbb{E}\lambda_n(t) - \lambda(t))^2 dt \\ &= \int_0^{T_F} \mathrm{Var}(\hat{E}\lambda_n(t))dt + \int_0^{T_F} (\mathbb{E}(\hat{E}\lambda_n(t) - \lambda(t))^2 dt + O(\frac{\log\log n}{n}). \qquad (4.16)\end{aligned}$$

Making use (3.2), we have

$$\begin{aligned}\mathrm{MISE}(\lambda_n) &= \frac{1}{n2^{-j}}\int_0^{T_F} \lambda^*(t)dt + \int_0^{T_F} (\mathbb{E}(\hat{E}\lambda_n(t)) - \lambda(t))^2 dt \\ &+ O(\frac{1}{n}) + O(\frac{\log\log n}{n}) \\ &= \int_0^{T_F}\left[\int_0^{T_F} \hat{K}_j(t,x)(\lambda(x) - \lambda(t))dx\right]^2 dt + O(\frac{1}{n2^{-j}}) + O(\frac{\log\log n}{n}). \qquad (4.17)\end{aligned}$$

Let $v_n(t) = \int_0^{T_F} \hat{K}_j(t,x)(\lambda(x) - \lambda(t))dx$.

As in the proof of Lemma 4.2 (see (4.9)), we obtain

$$\begin{aligned}|v_n(t)| &\le \int_{-2^j t}^{2^j(T_F - t)} \hat{K}(2^j t, 2^j t + s) \mid \lambda(t + 2^{-j}s) - \lambda(t) \mid ds \\ &\le C_0 \int_{-2L}^{2L} \hat{K}(2^j t, 2^j t + s) D_j(s) ds \qquad (4.18)\end{aligned}$$

where

$$D_j(s) = \mid f(t + 2^{-j}s) - f(t) \mid + \mid F(t + 2^{-j}s) - F(t) \mid.$$

Since $f^{(r+1)}$ exists and is bounded on $[0, T_F]$, Taylor's expansion of f and F up to r and the statements $(S1) - (S3)$ yield

$$\begin{aligned}|v_n(t)| &\le C_0 \frac{2^{-(r+1)j}}{(r+1)!}\int_{-2L}^{2L} |u|^{r+1} \hat{K}(2^j t, 2^j t + u) \times \\ &\left(\left|f^{(r+1)(t+\theta_1 2^{-j}u)}\right| + \left|f^{(r)(t+\theta_2 2^{-j}u)}\right|\right)du \le C_1 2^{-(r+1)j}, \qquad (4.19)\end{aligned}$$

where C_1 is a constant. Hence, by (4.17) we have

$$\text{MISE}(\lambda_n) = O(\frac{1}{n2^{-j}}) + O(2^{-(2r+2)j}) + O(\frac{\log\log n}{n}),$$

which is (3.4).

Proof of Theorem 3.3: We can write

$$\lambda_n(t) = \frac{1}{n}\sum_{i=1}^{n} \hat{K}_j(t, Z_i)\frac{\delta_i}{1-\hat{H}_n(Z_i)} = \int_0^{T_F} \hat{K}_j(t,x)\frac{d\tilde{H}_n(x)}{1-\hat{H}_n(x^-)} \tag{4.20}$$

where

$$\tilde{H}_n(x) = \frac{1}{n}\sum_{i=1}^{n} \delta_i 1_{\{Z_i \le x\}} = \frac{1}{n}\sum_{i=1}^{n} 1_{\{Z_i \le x, \delta_i = 1\}}. \tag{4.21}$$

$\tilde{H}_n$ is the empirical distribution function of the ith uncensored observation $(Z_i, \delta_i = 1)$ whose distribution function is

$$\tilde{H}(x) = \mathbb{P}[Z \le x,\ \delta = 1] = \int_0^x (1-G(s))dF(s).$$

Let

$$\Lambda(x) = \int_0^x \frac{dF(s)}{1-F(s^-)} = -\log(1-F(x)),$$

and

$$\Lambda_n(x) = \int_0^x \frac{d\tilde{H}_n(s)}{1-\hat{H}_n(s^-)} = -\log(1-\hat{F}_n(x)),\ x \le T_H.$$

Note that $\Lambda(x)$ and $\Lambda_n(x)$ denote the cumulative and Kaplan-Meier empirical cumulative hazard functions respectively. Since $d\tilde{H}(x) = (1-G(x))dF(x)$, a natural estimator of $d\tilde{H}(x)$ is $d\tilde{H}_n(x) = (1-\hat{G}_n(x))d\hat{F}_n(x)$. In view of (4.20), we readily have

$$\begin{aligned} \lambda_n(t) - \lambda(t) &= \int_0^{T_F} \hat{K}_j(t,x)\frac{d\tilde{H}_n(x)}{(1-\hat{H}_n(x^-))} - \int_0^{T_F} \hat{K}_j(t,x)\lambda(x)dx + O(2^{-j}) \\ &= \int_0^{T_F} \hat{K}_j(t,x)\frac{1-\hat{G}_n(x)}{1-\hat{H}_n(x^-)}d\hat{F}_n(x) - \int_0^{T_F} \hat{K}_j(t,x)d\Lambda(x) + O(2^{-j}) \\ &= \int_0^{T_F} \hat{K}_j(t,x)d\left(\Lambda_n(x) - \Lambda(x)\right) + O(2^{-j}). \end{aligned} \tag{4.22}$$

But

$$\begin{aligned} \int_0^{T_F} \hat{K}_j(t,x)d(\Lambda_n(x) - \Lambda(x)) &= 2^j \int_0^{T_F} \hat{K}(2^jt, 2^jx)d(\Lambda_n(x) - \Lambda(x)) \\ &= \Delta_j(t) - 2^{2j}\int_0^{T_F} (\Lambda_n(x) - \Lambda(x))\hat{K}'(2^jt, 2^jx)dx \end{aligned} \tag{4.23}$$

where

$$\Delta_j(t) = [(\Lambda_n(x) - \Lambda(x))\hat{K}(2^jt, 2^jx)]_0^{T_F}.$$

Making use of the statement (S2), we readily have $\hat{K}(2^jt, 2^jT_F) = 0$ for $0 \le t < T_F$.

Since $\Lambda_n(0) = \Lambda(0) = 0$, we see that $\Delta_j(t) = 0$. Hence, in view of (4.4) with $\hat{K}$ replaced by $\hat{K}'$, we obtain

$$
\begin{aligned}
|\lambda_n(t) - \lambda(t)| &\leq | 2^{2j} \int_0^{T_F} (\Lambda_n(x) - \Lambda(x))\hat{K}'(2^j t, 2^j x)dx | + O(2^{-j}) \\
&\leq 2^j \sup_x | \Lambda_n(x) - \Lambda(x) | \times C2^j \int_0^{T_F} \frac{1}{(1+2^j | x-t |)^2} dx + O(2^{-j}) \\
&\leq 2^j \sup_x | \Lambda_n(x) - \Lambda(x) | \frac{C'}{1+2^j(\text{dist}(t,[0,T_F]))^2} + O(2^{-j}). \qquad (4.24)
\end{aligned}
$$

Since $t \in [0, T_F]$, the expression in the right-hand side of (4.24) is less or equal to $C'2^j \sup_x | \Lambda_n(x) - \Lambda(x) | + O(2^{-j})$.

But, by Lemmas 2 and 4 of Diehl and Stute[7], $\sup_{0\leq t\leq T} | \Lambda_n(t) - \Lambda(t) |$ behaves like $O\Big(\sqrt{\frac{\log\log n}{n}}\Big)$. It follows from the last inequality that,

$$
\sup_{0\leq t\leq T} |\lambda_n(t) - \lambda(t)| \leq C'M2^j \Big(\frac{\log\log n}{n}\Big)^{\frac{1}{2}} + O(2^{-j}) \quad \text{a.s.} \qquad (4.25)
$$

where M is a constant. So, the Theorem 3.3 is proved.

Acknowledgments

The author thanks the referees for their careful reading of the manuscript and insightful comments. He also thanks the staff of LSTA of University of Paris 6 for their hospitality during his visit from September to December 2004 when this work was started.

References

[1] A. Antoniadis, G. Grégoire and G. Nason, Density and hazard rate estimation for right-censored data using wavelet methods. *J. R. Statist. Soc.* **61** (1999), pp. 63-84

[2] J. B. Aubin and A. Massiani, Comportement asymptotique d'un estimateur de la densité adaptatif par méthode d'ondelettes. *C.R. Acad. Sci. Paris ser I.* **337** (2003), pp. 293-296

[3] A. Carbonez, L. Gyöfi and E. C. Van der Meulen, Partition-estimates of a regression function under random censoring. *Statist. Decisions* **13** (1995),pp. 21-37

[4] S. Csörgő, Universal gaussian approximations under random censorship. *Ann. Statist.* **6** (1996), pp 2744-2778

[5] P. Deheuvels, Conditions nécessaires et suffisantes de convergence presque sûre et uniforme presque sûre des estimateurs de la densité. *C.R. Acad. Sci. Paris. Ser. A* **278** (1974), pp. 1217-1220

[6] P. Deheuvels and J. H. J. Einmhal, Functional limit laws for the increments of Kaplan-Meier product-limit processes and applications. *Ann. Probab.* **3** (2000), pp. 1301-1335

[7] S. Diehl and W. Stute, Kernel density and hazard function estimation in the presence of censoring. *J. Multivariate Anal.* **25** (1988), pp. 299-310

[8] A. Földes and L. Rejtő, A LIL type result for the product-limit estimator. *Z. Wahrsch. Verw. Gebiete* **56** (1981), pp. 775-86

[9] A. Földes, L. Rejtő and B. B. Winter, Strong consistency properties of nonparametric estimators for randomly censored data II: the estimation of density and failure rate. *Period. Math. Hungar* **12** (1981), pp. 15-29

[10] K. E. Gneyou, *Inférence statistique pour l'analyse du taux de panne en fiabilité* . Thèse de Doctorat de l'Université Paris VI (1991) pp. 1-150

[11] K. E. Gneyou, Normalit asymptotique d'une fonctionnelle du taux de panne bas sur des donnes avec censures alatoire droite. *Ann. Uni. Bénin, série Sciences*, Tome XII, (1996) pp. 3-15

[12] K. E. Gneyou, A functional law of iterated logarithm for a hazard rate process from censored data. *J. Rech. Sci. Uni. Bénin (Togo),* **1** (2) (1997), pp. 44-48

[13] K. E. Gneyou, Vitesse de convergence de certains estimateurs de Kapla-Meier de la régrssion. *Afrika Statistika,* **1** (1) (2005) pp. 77-92

[14] M. G. Gu and T. L. Lai, Functional laws of the iterated logarithm for the product-limit estimator of a distribution function under random censorship or truncation. *Ann. Probab.* **18** (1990), pp. 160-189

[15] P. Hall and P. Patil, Formula for mean integrated squared error of nonlinear wavelet-based density estimators. *Ann. Statist.* **23** (1995), pp. 905-928

[16] W. Härdle, G. Kerkyacharian, D. Picard and A. Tsybakov, *Wavelets, approximation, and statistical applications*. Springer-Verlag, New York,(1998)

[17] I. Johstone, G. Kerkyacharian and D. PICARD, Estimation d'une densité de probabilité par méthodes d'ondelette. *C.R. Acad. Sci. Paris* **315** (1992), pp. 211-216

[18] E. K. Kaplan and P. Meier, *Nonparametric estimation from incomplete observations* . *J. Amer. Statist. Assoc.*, **53** (1958), pp 457-481.

[19] G. Kerkyacharian and D. Picard, Estimation de densité par méthodes de noyau et d'ondelette: les liens entre la géométrie du noyau et les contraintes de régularité. *C.R. Acad. Sci. Paris t315, ser I* (1992), pp 79-84

[20] G. Kerkyacharian and D. Picard, Density estimation in Besov space. *Statist. and Probab. Letters* **13** (1992), pp 15-24

[21] S. Mallat, Multiresolution approximations and wavelet orthonormal bases of $L^2(\mathbb{R})$. *Trans. Amer. Math. Soc.* **315** (1989), pp 69-87

[22] Y. Meyer *Ondelettes et operateurs I*, Hermann, Paris 1990, 315 (1990), pp 69-87

[23] J. Mielniczuk, Some asymptotic properties of kernel estimators of a density function in case of censored data. *Ann. Statist.* **1** (1986), pp 766-773

[24] H. G. Müler and L. L. Wang, Hazard rate estimation under random censoring varying kernels and bandwidths. *Biometrics.* **50** (1994), pp 61-76

[25] W. Stute, A law of the iterated logarithm for kernel density estimators. *Ann. Probab.* **10** (1982b), pp 414-422

[26] M. A. Tanner and W. H. Wong, The estimation of the hazard function from randomly censored data by the kernel method. *Ann. Statist.* **11** (1983), pp 983-993

[27] G. S. Watson and M. R. Leadbetter, Hazard analysis I. *Biometrika* **51** (1964a), pp 175-184

[28] G. S. Watson and M. R. Leadbetter, Hazard analysis II. *Sankhyà, Ser. A* **26** (1964b), pp 101-116

In: African Diaspora Mathematics Research Progress ISBN 978-1-60456-204-0
Editor: Toka Diagana, pp. 105–128 © 2008 Nova Science Publishers, Inc.

Chapter 8

SINGULAR REDUCTION AND STRATIFICATION OF QUIVER VARIETY

Bassirou Diatta*
Department of Mathematics, Jackson State University,
P.O. Box 17610, Jackson, MS, 39217-0410, USA
Stanley M. Einstein-Matthews†
Department of Mathematics, Howard University,
2441 6th Street N.W., Washington, DC 20059, USA

Abstract

In this article we study singular reduction and stratification in the case of the action of a complex reductive Lie group on a Quiver Variety. The main result of the paper is an illustration of the key role R. Sjamaar's Holomorphic Slice Theorem can play in the understanding of some interesting aspects of singular reduction theory.

AMS Subject Classification: Primary 58FA40, Secondary 58FA05.

Keywords: complex structure, differential space, symplectic space, Poisson reduction, singular reduction, stratified space.

1 Introduction

The goal in this paper is to use the representation theory of quivers

$$Q = (Q_0, Q_1, s, t : Q_1 \longrightarrow Q_0)$$

of finite type with finite set of vertices $Q_0 := \{1, 2, \ldots, n\}$ and finite set of arrows Q_1 together with functions $s, t : Q_1 \longrightarrow Q_0$ to study the action of the complex reductive algebraic Lie group $G = \bigoplus_{k \in Q_0} GL_C(V_k)$, on the representation spaces $(V, \varphi) = (V_k, \varphi_h)_{k \in Q_0, h \in Q_1}$, to obtain the quotient space $Rep^{\lambda}_{\overline{Q}}(v, w)$, which we call a quiver variety in line with the general

*E-mail: basirou.diatta@jsums.edu, bdiatta@howard.edu
†E-mail: seinatein-matth@howard.edu

trend in the mathematical literature (see [4], [17], [12]). We then employ the general stratification theory developed in ([21], [22]), to understand the symplectic smooth strata of stratified singular quiver varieties together with the holomorphic stratification of semistable and stable subspaces of $Rep_{\overline{Q}}(v,w)$ considered as Kähler and hyper-Kähler manifolds. The main idea is to show how R. Sjamaar's Holomorphic Slice Theorem can be used as an important tool to understand the reduction problem for stratifications of the quiver variety $Rep_{\overline{Q}}^{\lambda}(v,w)$. It is worth mentioning here that $Rep_{\overline{Q}}^{\lambda}(v,w)$ carries the three stratifications: Hyper-Kähler stratification as in Hitchin et all [7], holomorphic stratification (cf. R. Sjamaar [22]), and symplectic stratification, (cf. R. Sjamaar and E. Lerman [23]). In the sequel we construct smooth structure on the reduced space $Rep_{\overline{Q}}^{\lambda}(v,w)$. The Theorems 2.8 and 2.9 of [22], equip $Rep_{\overline{Q}}^{\lambda}(v,w)$ with a ring of complex analytic structures thus turning it into a ringed space and exhibiting its precise affine algebraic variety character. It is then clear that a choice of Hermitian metrics on the representation spaces $(V,\varphi)=(V_k,\varphi_h)_{k\in Q_0,h\in Q_1}$ of the quivers $Q=(Q_0,Q_1,s,t:Q_1\longrightarrow Q_0)$ can be used to see that $Rep_{\overline{Q}}^{\lambda}(v,w)$ is actually a flat hyper-Kähler space and hence deduce that: Holomorphic symplectic reduction is equivalent to hyper-Kähler reduction. Furthermore, we observe that for a fixed $\lambda\in\mathcal{G}^*\simeq\mathcal{G}=End(V)$, and for all Ad^*-equivariant orbits $\mathcal{O}_\lambda\subset\mathcal{G}^*$ through λ, there exists a hyper-Kähler structure on $Rep_{\overline{Q}}^{\lambda}(v,w)$ as shown in (Pflaum [21], R. Sjamaar and E. Lerman [23], H. Nakajima ([17], [18], [19])). We further show that $Rep_{\overline{Q}}^{\lambda}(v,w)$ admits a Poisson structure as a stratified space since the stratified pieces are smooth. Our main result, the Singular Reduction Theorem, is an essential application of R. Sjamaar's Holomorphic Slice Theorem. In addition we exhibit Hamiltonian dynamics on the reduced symplectic space.

We conclude this introduction with a brief description of the layout of this paper. Apart from the discussion of the motivation in section one, Section two is devoted to setting up our notation , statement of relevant facts and definitions used in the main body of the paper. This section also contains important constructions and proofs of the theorems we use in the final section. In section three we state the main results of the paper and explain the important geometric constructions from section two and their relations to the proof of our main theorem.

2 Preliminaries and Background

Fix $\mathbb{K}$ an algebraically closed field of characteristic zero, which is either $\mathbb{K}=\mathbb{R}$ or $\mathbb{C}$. Let (Q_0,Q_1) be a connected finite graph, with $Q_0:=\{1,2,\cdots,n\}$ the set of vertices of cardinality n and Q_1 the set of arrows of the graph and the orientation is given by two maps:

$$\begin{array}{rccl} s,\ t\ : & Q_1 & \longrightarrow & Q_0 \\ & h & \mapsto & s(h) \text{ the source map, and} \\ & h & \mapsto & t(h) \text{ the target map} \end{array} \tag{2.1}$$

such that

1. $\forall h\in Q_1,\ s(h)\neq t(h)$.
2. An involution $\sigma:Q_1\longrightarrow Q_1;\ h\mapsto\sigma(h)=\bar{h}$, without fixed points satisfying $s(\bar{h})=t(h)$, is fixed,

3. A map $\varepsilon : Q_1 \longrightarrow \{1, -1\}$ given by $\varepsilon(\bar{h}) = -\varepsilon(h)$.

Now define the sets

$$\Omega := \{h \in Q_1 \ : \ \varepsilon(h) = 1\} \text{ and}$$
$$\overline{\Omega} := \{h \in Q_1 \ : \ \varepsilon(h) = -1\}.$$

We can then associate to every finite connected graph without edge loop (Q_0, Q_1) a generalized symmetric Cartan matrix C. To do this we let A_Q be the matrix of the quiver Q, which is the $Q_0 \times Q_0$ matrix defined by

$$A_Q(l,k) := \#\{h \in Q_1 : s(h) = k, t(h) = l\}$$

with entries the numbers,

$$a_{kl} := \{h \in Q_1 \ : \ s(h) = k, \text{ and } t(h) = l\}.$$

Define the generalized symmetric Cartan matrix $C = (c_{kl})_{k,l \in Q_0}$ by setting $C = 2I - A_Q$ with $I = (I_{ij})_{i,j \in Q_0}$ the identity matrix. The cardinality of $C = (c_{kl})$ given by Cardc_{kl}, is the number of arrows joining k and l in Q_0 in this order, so that

$$c_{kl} = \begin{cases} -c_{lk}, & if\, k \neq l \\ 2, & if\, k = l \end{cases}$$

This establishes a bijection between finite graphs without loops and generalized symmetric Cartan matrices. The sets Ω, $\overline{\Omega} \subset Q_1$ are such that

$$\Omega \cap \overline{\Omega} = \varnothing \text{ and } Q_1 = \Omega \cup \overline{\Omega}.$$

The orientation of the graph (Q_0, Q_1) as defined can now be seen as represented by $\Omega \subset Q_1$ and the maps

$$\begin{array}{cccc} s,\ t\ : & Q_1 & \longrightarrow & Q_0 \\ & h & \mapsto & s(h) \\ & h & \mapsto & t(h) \end{array} \tag{2.2}$$

are the source of $h \in Q_1$, and $t(h)$ target of $h \in Q_1$ respectively; such that $s(h) = t(\bar{h})$, $s(\bar{h}) = t(h)$ for some h, $\bar{h} \in Q_1$. Define the quiver $\overline{Q}$; the double of the quiver Q; as the quiver obtained from Q by adjoining a reverse arrow $y^* := \bar{y} : \overset{l}{\bullet} \longrightarrow \overset{k}{\bullet}$ for each arrow $y : \overset{k}{\bullet} \longrightarrow \overset{l}{\bullet}$ in Q_1, i.e.,

$$\overset{k}{\bullet} \ \underset{y}{\overset{y^*}{\leftrightarrows}} \ \overset{l}{\bullet} \tag{2.3}$$

Let $V = (V_k)_{k \in Q_0}$ and $W = (W_k)_{k \in Q_0}$ be two families of finite dimensional $\mathbb{K}$-vector spaces. Define their dimension vectors v and w by:

$$v := (dimV_k)_{k \in Q_0} \ = \ (dimV_1, \cdots, dimV_n) \in \mathbb{Z}^{Q_0}_{\geq 0}$$

and

$$w := (dimW_k)_{k \in Q_0} \ = \ (dimW_1, \cdots, dimW_n) \in \mathbb{Z}^{Q_0}_{\geq 0}.$$

At each vertex $k \in Q_0$, we attach finite dimensional $\mathbb{K}$-vector spaces V_k and W_k.

Definition 2.1 *A representation* (V,φ) *of a quiver* Q *over the field* $\mathbb{K}$ *is a family of finite dimensional* $\mathbb{K}$*-vector spaces* $V = (V_k)_{k\in Q_0}$ *together with* $\mathbb{K}$*-linear maps*

$$\varphi_h : V_{k=s(h)} \longrightarrow V_{l=t(h)},$$

for each arrow $h : \overset{k}{\bullet} \longrightarrow \overset{l}{\bullet}$ *in* Q_1.

Next we borrow some notations from V. Ginzburg [4], H. Nakajima ([17], [18], [19]), and adapt them to suite our purposes. We define the following finite dimensional $\mathbb{C}$-vector spaces

$$Rep_\Omega(v,\, w) := \oplus_{h\in\Omega} Hom_{\mathbb{C}}(V_{s(h)},\, V_{t(h)}) \oplus (\oplus_{l\in Q_0} Hom_{\mathbb{C}}(W_l,\, V_l)),$$

and

$$Rep_{\overline{\Omega}}(v,\, w) := \oplus_{\bar{h}\in\overline{\Omega}} Hom_{\mathbb{C}}(V_{s(\bar{h})},\, V_{t(\bar{h})}) \oplus (\oplus_{l\in Q_0} Hom_{\mathbb{C}}(V_l,\, W_l)).$$

Then we set

$$\begin{aligned} Rep_{\overline{Q}}(v,\, w) &:= Rep_\Omega(v,\, w) \oplus Rep_{\overline{\Omega}}(v,\, w) \\ &:= T^\star Rep_\Omega(v,\, w) \end{aligned}$$

a finite dimensional $\mathbb{C}$- vector space, it is the cotangent bundle of $\mathbb{C}$-vector space $Rep_\Omega(v,\, w)$.
By the dual pairing, let $\delta \in Hom_{\mathbb{C}}(V_l, W_l)$ and $\gamma \in Hom_{\mathbb{C}}(W_l, V_l)$ for $l \in Q_0$, and define $<\delta,\gamma>:= trace(\delta\circ\gamma)$. Define the real dimension of $Rep_{\overline{Q}}(v,\, w)$ by

$$\begin{aligned} \dim_{\mathbb{R}} Rep_{\overline{Q}}(v,\, w) &= \Sigma_{k,l=1}^{n} 2a_{kl}v_kv_l + 2wv^\tau + 2wv^\tau \\ &= 2vA_Qv^\tau + 4wv^\tau, \end{aligned} \tag{2.4}$$

where v^τ denotes the transpose of the dimension vector $v \in \mathbb{Z}^{Q_0}_{\geq 0}$ and the sum is taken over all arrows of $\overline{Q}$ so that each arrow appears twice in the sum, once for each direction, $A_Q := (a_{kl})_{k,l\in Q_0}$ is the symmetrizable generalized Cartan matrix which is the adjacency matrix of the graph (Q_0, Q_1). The number of arrows $h \in Q_1$ joining the vertices $k,l \in Q_0$ in this order, i.e., $k = s(h) \longrightarrow l = t(h)$ is defined as a cardc_{kl}. We define a multiplication in $Rep_{\overline{Q}}(v,w)$ by a composition operations see Nakajima [17].
Let $(\varepsilon\varphi)_h := \varepsilon(h)\varphi_h$, $\forall h \in Q_1$, (φ,a,b), and (φ',a',b') be elements in $Rep_{\overline{Q}}(v,\, w)$, such that the skew-symmetric bilinear closed 2-form is defined by

$$\omega((\varphi,a,b),(\varphi',a',b')) := trace(\varepsilon\varphi\varphi') + trace(ab' - a'b).$$

The form ω is a symplectic form and the pair $(Rep_{\overline{Q}}(v,w),\omega)$ is a symplectic $\mathbb{C}$-vector space and hence a symplectic manifold.
Let $G = \prod_{k\in Q_0} GL_{\mathbb{C}}(V_k)$ be a complex reductive Lie group and $K := \prod_{k\in Q_0} U(V_k)$ a maximal compact real Lie subgroup of G. We view the complex reductive Lie group G as the complexification $K^{\mathbb{C}} = K \otimes_{\mathbb{R}} \mathbb{C}$,of K. Let $\mathcal{K}$ =Lie(K), be Lie algebra of K and $\mathcal{G}$ the Lie algebra of G. Suppose the K-action on the complex symplectic manifold $(Rep_{\overline{Q}}(v,w),\omega)$:

$$\Psi_K : K \times Rep_{\overline{Q}}(v,w) \longrightarrow Rep_{\overline{Q}}(v,w)$$

is proper, Hamiltonian and has an Ad^*- coadjoint equivariant momentum map

$$\Phi_K : Rep_{\overline{Q}}(v,w) \longrightarrow \mathcal{K}^*,$$

where $\mathcal{K}^*$ is the dual of the Lie algebra $\mathcal{K} = \mathrm{Lie}(K)$, with the property that $d\Phi_K^{\xi} = i_{\xi_{Rep_{\overline{Q}}(v,w)}}\omega$, for all $\xi \in \mathcal{K}$ and the vector field $\xi_{Rep_{\overline{Q}}(v,w)}$ on $Rep_{\overline{Q}}(v,w)$ induced by $\xi \in \mathcal{K}$ is Hamiltonian. The induced map Φ_K^{ξ} is the ξth component of Φ_K defined by $\Phi_K^{\xi}((\varphi,a,b)) := (\Phi_K(\varphi,a,b))(\xi)$.

The Hamiltonian action of G on $Rep_{\overline{Q}}(v,w)$:

$$\begin{array}{ccc} G \times Rep_{\overline{Q}}(v,w) & \stackrel{\Psi}{\longrightarrow} & Rep_{\overline{Q}}(v,w) \\ (g,(\varphi,a,b)) & \mapsto & (g\varphi g^{-1}, ga, bg^{-1}) \end{array}$$

is given by conjugation and leaves invariant the symplectic form ω on $Rep_{\overline{Q}}(v,w)$ and induces an Ad^* -equivariant momentum map

$$\begin{array}{rl} \Phi: & Rep_{\overline{Q}}(v,w) \longrightarrow \mathcal{G}^* \\ & (\varphi,a,b) \mapsto \Phi(\varphi,a,b) := \varepsilon\varphi\varphi + ab, \end{array}$$

where $\mathcal{G}^*$ is dual of the Lie algebra $\mathcal{G} := Lie(G)$. Next assume the value of the momentum map is fixed by the coadjoint action Ad^* on $\mathcal{G}^*$ such that the following diagram

$$\begin{array}{ccc} G \times Rep_{\overline{Q}}(v,w) & \stackrel{id_G \times \Phi}{\longrightarrow} & G \times \mathcal{G}^* \\ \Psi \downarrow & & \downarrow Ad_G^* \\ Rep_{\overline{Q}}(v,w) & \stackrel{\Phi}{\longrightarrow} & \mathcal{G}^* \end{array} \tag{2.5}$$

commutes preserving the momentum map Φ.

Let $\lambda \in \mathcal{G}^*$. Then consider all closed co-adjoint orbits $O_\lambda \subset \mathcal{G}^*$ passing through the point λ, and all closed co-adjoint orbits $O_{\lambda_{\mathcal{K}^*}} \subset \mathcal{K}^*$, where $\lambda_{\mathcal{K}^*} \in \mathcal{K}^*$ is the real component of $\lambda \in \mathcal{G}^*$. Identify $\mathcal{G}^*$, the dual of the Lie algebra $\mathcal{G} := \mathrm{Lie}(G)$ of G by using the trace pairing:

$$\begin{array}{rccc} trace: & \mathcal{G} \times \mathcal{G}^* & \longrightarrow & \mathbb{R} \\ & (\xi,\sigma) & \mapsto & \Sigma_{k\in Q_0} trace(\xi_k,\sigma_k). \end{array}$$

Set

$\mathcal{G}^* \simeq subal(\mathcal{G}) := \{\xi = (\xi_k)_{k\in Q_0} \in \Sigma_{k\in Q_0} \mathcal{G}l(V_k) : \Sigma_{k\in Q_0} trace(\xi_k) = 0\}$. Then by the use of geometric invariant theory define the quotient

$$\begin{aligned} Rep_{\overline{Q}}^{\lambda_{\mathcal{K}^*}}(v,w) &:= \Phi^{-1}(O_{\lambda_{\mathcal{K}^*}})//K \\ &= Spec(\mathbb{C}[Rep_{\overline{Q}}(v,w)])^K / I(\Phi^{-1}(O_{\lambda_{\mathcal{K}^*}}))^K, \end{aligned}$$

where $I(\Phi^{-1}(O_{\lambda_{\mathcal{K}^*}})) := \{f \in \mathbb{C}[Rep_{\overline{Q}}(v,w)];\ f|_{\Phi^{-1}(O_{\lambda_{\mathcal{K}^*}})} = 0\}$ is the defining ideal of $\Phi^{-1}(O_{\lambda_{\mathcal{K}^*}})$.

Similarly, define the quotient

$$\begin{aligned} Rep_{\overline{Q}}^{\lambda}(v,w) &:= \Phi^{-1}(O_\lambda)//G \\ &= Spec(\mathbb{C}[Rep_{\overline{Q}}(v,w)]^G / I(\Phi^{-1}(O_\lambda))^G), \end{aligned}$$

where $I(\Phi^{-1}(O_\lambda)) := \{f \in \mathbb{C}[Rep_{\overline{Q}}(v,w)];\ f|_{\Phi^{-1}(O_\lambda)} = 0\}$ is the defining ideal of $\Phi^{-1}(O_\lambda)$.

Definition 2.2 *The quotients $Rep_{\overline{Q}}^{\lambda_{\mathcal{K}^*}}(v,w)$ and $Rep_{\overline{Q}}^{\lambda}(v,w)$ are algebraic varieties, called affine quiver varieties or simply quiver varieties in the sequel.*

Remark 2.3

1. *If, $\lambda \in O_{\lambda_{\mathcal{K}^*}} \subset \mathcal{K}^*$ and $\lambda \in O_\lambda \subset \mathcal{G}^*$ are regular, then the quiver varieties $Rep_{\overline{Q}}^{\lambda_{\mathcal{K}^*}}(v,w)$ and $Rep_{\overline{Q}}^{\lambda}(v,w)$ are manifolds [4], otherwise*
2. *These quiver varieties $Rep_{\overline{Q}}^{\lambda_{\mathcal{K}^*}}(v,w)$ and $Rep_{\overline{Q}}^{\lambda}(v,w)$ are not manifolds if $\lambda_{\mathcal{K}^*}$ and λ are singular, but they have very interesting properties such as stratifications, smooth structures on the strata, stability properties.*

We now recall from H. Nakajima ([17], [18], [19]) and King [9], the following definition for the stability of points of the quiver varieties $Rep_{\overline{Q}}^{\lambda_{\mathcal{K}^*}}(v,w)$ and $Rep_{\overline{Q}}^{\lambda}(v,w)$.

Definition 2.4 *(H. Nakajima, [19] definition 2.3.1) A point $(\varphi,a,b) \in \Phi^{-1}(O_\lambda)$ is said to be a stable point if the following conditions hold*

If a collection $S = (S_k)_{k\in Q_0}$ of subspaces of $V = (V_k)_{k\in Q_0}$ is φ- invariant and contained in Kerb, then $S = 0$.

lift the action of G on $\Phi^{-1}(O_\lambda)$ to the trivial line bundle $\Phi^{-1}(O_\lambda) \times \mathbb{C}$ defined by $g(\varphi,a,b,z) := (g(\varphi,a,b),\chi(g)^{-1}z)$, where $\chi : K \longrightarrow \mathbb{K}^\star = \mathbb{K}\backslash\{0\}$ for each $g \mapsto \chi(g) = \prod_{l\in Q_0} det(g_l^{-1})$ is the character map for any $g = (g_l)_{l\in Q_0}$

and define $\Phi^{-1}(O_\lambda)^s := \{(\varphi,a,b) \in \Phi^{-1}(O_\lambda) :\ \overline{G(\varphi,a,b)} \cap (\Phi^{-1}(O_\lambda) \times \{0\}) = \emptyset \text{ for } z \neq 0\}$ the set of all stable points of $\Phi^{-1}(O_\lambda)$. H. Nakajima has shown also that the semistable and stable points coincide in this case. The following result of H. Nakajima ([17], Lemma 3.10) gives an important observation about stable points.

Proposition 2.5 *Let $\Phi^{-1}(O_\lambda) \subset Rep_{\overline{Q}}(v,w)$ and a point $(\varphi,a,b) \in \Phi^{-1}(O_\lambda)$ such that $(\varphi,a,b) \in \Phi^{-1}(O_\lambda)^s$ is a stable point. Then*

1. *The stabilizer of (φ,a,b) in G is trivial.*
2. *The differential operator $d\Phi : Rep_{\overline{Q}}(v,w) \longrightarrow \mathcal{G}^*$ is surjective at (φ,a,b), thus $\Phi^{-1}(O_\lambda)^s$ is a nonsingular algebraic subvariety of dimension $v^\tau(2w+(I-C)v)$ where I is the identity matrix and $C = (c_{kl})_{k,l\in Q_0}$ is the symmetric generalized Cartan matrix.*

Proof. See H. Nakajima [17]. We give an idea of the proof here for ease of reference.

1. Let $g = (g_l)_{l\in Q_0} \in G$ stabilize a point $(\varphi,a,b) \in \Phi^{-1}(O_\lambda)$. The family of subspaces $Im(g_l^{-1})$ is then φ-invariant and lies in the kernel of the $\mathbb{C}$-linear maps $b_l : V_l \longrightarrow W_l$, $\forall l \in Q_0$. Stability condition then implies $g_l = 1$, $\forall l \in Q_0$.
2. Pick $\alpha \in \mathcal{G}^*$ orthogonal to the $Im\{d\Phi : Rep_{\overline{Q}}(v,w) \longrightarrow \mathcal{G}^*\}$ with respect to inner product given by the trace identifying $\mathcal{G}$ and $\mathcal{G}^*$, so that $\mathcal{G} = End(V)$. Then deduce that $\varphi\alpha = \alpha\varphi$, $\alpha a = 0$ and $b\alpha = 0$. Thus $Im\alpha$ is φ-invariant and is contained in $ker\{b : V \longrightarrow W\}$. Stability condition once again shows that $\alpha = 0$. ◇

From here onwards we will restrict our discussions to the singular quiver varieties and start by first examining the structures of the quiver variety $Rep_{\overline{Q}}^{\lambda_{\mathcal{K}^*}}(v,w)$. Sjamaar and Lermann [23], and also H. Nakajima [18] have studied the action of the compact Lie group K on the symplectic manifold $Rep_{\overline{Q}}(v,w)$.
For any point $(\varphi_0,a_0,b_0)\in Rep_{\overline{Q}}(v,w)$, denote by $H:=K_{(\varphi_0,a_0,b_0)}=\{k\in K:\ k(\varphi_0,a_0,b_0)=(\varphi_0,a_0,b_0)\}\subset K$ the stabilizer of (φ_0,a_0,b_0) in K, and define by

$$Rep_{\overline{Q}}(v,w)_{(H)}:=\{(\varphi,a,b)\in Rep_{\overline{Q}}(v,w):\ \ kK_{(\varphi,a,b)}k^{-1}=H,\ \ \forall k\in K\},$$

all the points of orbit type (H), where (H) is the conjugacy class of the subgroup H. This is a symplectic submanifold of $Rep_{\overline{Q}}(v,w)$ with its symplectic structure $\omega_{Rep_{\overline{Q}}(v,w)_{(H)}}:=\omega|_{Rep_{\overline{Q}}(v,w)_{(H)}}$ which is the restriction of ω to $Rep_{\overline{Q}}(v,w)_{(H)}$. Then $Rep_{\overline{Q}}(v,w)$ is a stratified symplectic manifold which is a disjoint union of symplectic submanifolds called the strata satisfying suitable criteria. Write

$$Rep_{\overline{Q}}(v,w)=\bigsqcup_{(H)}Rep_{\overline{Q}}(v,w)_{(H)},$$

(see M. Goresky and R. MacPherson [5], and M.J. Pflaum [21]). A point $[(\varphi,a,b)]\in Rep_{\overline{Q}}^{\lambda_{\mathcal{K}^*}}(v,w)$ is said to be of H-orbit type (H) if its representative (φ,a,b) is in $Rep_{\overline{Q}}(v,w)_{(H)}$. Define

$$Rep_{\overline{Q}}^{\lambda_{\mathcal{K}^*}}(v,w)_{(H)}:=Rep_{\overline{Q}}(v,w)_{(H)}\cap(\Phi^{-1}(O_{\lambda_{\mathcal{K}^*}})//K),$$

to be the stratum of all the points of orbit type (H). The stratum $Rep_{\overline{Q}}^{\lambda_{\mathcal{K}^*}}(v,w)_{(\{e_K\})}$ is the stratum corresponding to the conjugacy class of the trivial subgroup $\{e_K\}\subset K$ of the identity element $e_K\in K$. From([23], [21], [17]), we know that the stratum $Rep_{\overline{Q}}^{\lambda_{\mathcal{K}^*}}(v,w)_{(H)}$ is a symplectic manifold, with its symplectic structure

$$\pi^*\omega_{Rep_{\overline{Q}}^{\lambda_{\mathcal{K}^*}}(v,w)_{(H)}}=\omega_{Rep_{\overline{Q}}(v,w)_{(H)}},$$

where $\pi:Rep_{\overline{Q}}(v,w)\longrightarrow Rep_{\overline{Q}}^{\lambda_{\mathcal{K}^*}}(v,w)$ is the canonical projection map. From the above definitions, we deduce the following decomposition

$$Rep_{\overline{Q}}^{\lambda_{\mathcal{K}^*}}(v,w)=\bigsqcup_{(H)}Rep_{\overline{Q}}^{\lambda_{\mathcal{K}^*}}(v,w)_{(H)},$$

where the disjoint union is taken over all the set of the conjugacy classes of subgroups H of the group K.
R. Sjamaar and E. Lermann have shown in [23] that by defining the algebra of all smooth functions

$$C^{\infty}(Rep_{\overline{Q}}(v,w))=\{f:Rep_{\overline{Q}}(v,w)\longrightarrow\mathbb{R}\}$$

on the symplectic manifold $Rep_{\overline{Q}}(v,w)$, one can define the algebra of smooth functions on its stratum $Rep_{\overline{Q}}(v,w)_{(H)}$, by simply requiring that any function

$$h\in C^{\infty}(Rep_{\overline{Q}}(v,w)_{(H)})=\{h:Rep_{\overline{Q}}(v,w)_{(H)}\longrightarrow\mathbb{R}\},$$

is the restriction of a function $f \in C^\infty(Rep_{\overline{Q}}(v,w))$ on $Rep_{\overline{Q}}(v,w)_{(H)}$, denoted by

$$h := f|_{Rep_{\overline{Q}}(v,w)_{(H)}}.$$

We also define smooth functions

$$C^\infty(Rep_{\overline{Q}}^{\lambda_{\mathcal{K}^*}}(v,w)_{(H)}) = \{f : Rep_{\overline{Q}}^{\lambda_{\mathcal{K}^*}}(v,w)_{(H)} \longrightarrow \mathbb{R}\}$$

on the reduced stratum $Rep_{\overline{Q}}^{\lambda_{\mathcal{K}^*}}(v,w)_{(H)})$ in the same manner by requiring that for any $h \in C^\infty(Rep_{\overline{Q}}^{\lambda_{\mathcal{K}^*}}(v,w)_{(H)})$ there exists $f \in C^\infty(Rep_{\overline{Q}}(v,w))$ such that

$$\pi^* h := f|_{Rep_{\overline{Q}}(v,w)_{(H)}}.$$

Then

$$C^\infty(Rep_{\overline{Q}}^{\lambda_{\mathcal{K}^*}}(v,w)) := C^\infty(Rep_{\overline{Q}}(v,w))^K / _{I(\Phi^{-1}(O_{\lambda_{\mathcal{K}^*}}))},$$

where $C^\infty(Rep_{\overline{Q}}(v,w))^K$ is the algebra of K-invariant smooth functions on $Rep_{\overline{Q}}(v,w)$ and $I(\Phi^{-1}(O_{\lambda_{\mathcal{K}^*}})) := \{f \in C^\infty(Rep_{\overline{Q}}(v,w))^K : \ f|_{\Phi^{-1}(O_{\lambda_{\mathcal{K}^*}})} = 0\}$.
Let $[0,1]$ be an interval in $\mathbb{R}$, a smooth curve $c : [0,1] \longrightarrow Rep_{\overline{Q}}(v,w)_{(H)}$ on the smooth stratum $Rep_{\overline{Q}}(v,w)_{(H)}$ is an integral curve of a vector field $X_f \in Der(C^\infty(Rep_{\overline{Q}}(v,w)_{(H)}))$ through (φ_0, a_0, b_0) if

$$(X_f.f)(c(t)) = \frac{d}{dt} f(c(t))$$

for every $f \in C^\infty(Rep_{\overline{Q}}(v,w)_{(H)})$, $t \in [0,1]$, and $c(0) = (\varphi_0, a_0, b_0)$. It can be easily seen that the Poisson structure on the reduced space $Rep_{\overline{Q}}^{\lambda_{\mathcal{K}^*}}(v,w)$ is well-defined. Let $\{,\}_{\lambda_{\mathcal{K}^*}}$ be the Poisson structure on $Rep_{\overline{Q}}^{\lambda_{\mathcal{K}^*}}(v,w)$.

Definition 2.6 *(L. Bates, E. Lermann [1], definition 4)*
A Hamiltonian flow of a smooth function f on a reduced space $Rep_{\overline{Q}}^{\lambda_{\mathcal{K}^*}}(v,w)$ is a smooth flow $\{\phi_s\}_{s\in\mathbb{R}}$ such that for any point $[(\varphi,a,b)]$ in $Rep_{\overline{Q}}^{\lambda_{\mathcal{K}^*}}(v,w)$ and any smooth function $h \in C^\infty(Rep_{\overline{Q}}^{\lambda_{\mathcal{K}^*}}(v,w))$ we have

$$\frac{d}{ds} h(\phi_s)([(\varphi,a,b)]) = \{f,h\}(\phi_s([(\varphi,a,b))]), \tag{2.6}$$

where $\{,\}$ is the Poisson bracket on $C^\infty(Rep_{\overline{Q}}^{\lambda_{\mathcal{K}^*}}(v,w)$.

The fact that the reduced space $Rep_{\overline{Q}}^{\lambda_{\mathcal{K}^*}}(v,w)$ is not locally Euclidean, raises some difficulties in proving existence and uniqueness results for solutions to equation (2.6). However, the following lemma gives requisite conditions for overcoming these difficulties.

Lemma 2.7 *Let $Rep_{\overline{Q}}^{\lambda}(v,w)$ be the quiver variety. If the smooth functions $C^\infty(Rep_{\overline{Q}}^{\lambda}(v,w),\mathbb{C})$ on $Rep_{\overline{Q}}^{\lambda}(v,w)$ separate points, then the Hamiltonian flows $\{\varphi_s\}_{s\in\mathbb{R}}$ defined above are unique.*

Proof. The proof for this lemma is a simple adaptation of that in (E. Lerman and R. Sjamaar [23]). We refer the reader to the cited reference for details. The idea of the proof is as follows:
Consider two Hamiltonian flows φ_t, $\tilde{\varphi}_t$ on $Rep^{\lambda}_{\overline{Q}}(v,w)$ given by a function $F_\lambda \in C^\infty(Rep^{\lambda}_{\overline{Q}}(v,w),\mathbb{C})$. Then chain rule implies φ_{-t} is a flow for the negative function $-F_\lambda$. Because functions in $C^\infty(Rep^{\lambda}_{\overline{Q}}(v,w),\mathbb{C})$ separate points in $Rep^{\lambda}_{\overline{Q}}(v,w)$ it suffices to prove that for any $H_\lambda \in C^\infty(Rep^{\lambda}_{\overline{Q}}(v,w),\mathbb{C})$ and any point $m = [(\varphi,a,b)] \in Rep^{\lambda}_{\overline{Q}}(v,w)$ we have $H_\lambda(\tilde{\varphi}_s(\varphi_{-s}(m))) = H_\lambda(m)$. This follows from the following computation.

$$\frac{dH_{\lambda_{\mathcal{K}^*}}(\tilde{\varphi}_s(\varphi_{-s}(m))}{ds} =$$
$$= \{H_{\lambda_{\mathcal{K}^*}}, F_{\lambda_{\mathcal{K}^*}}\}_{\lambda_{\mathcal{K}^*}}(\tilde{\varphi}_s(\varphi_{-s}(m)) + \{H_{\lambda_{\mathcal{K}^*}}, -F_{\lambda_{\mathcal{K}^*}}\}_{\lambda_{\mathcal{K}^*}}(\tilde{\varphi}_s(\varphi_{-s}(m)) = 0. \quad \Diamond$$

Next, consider the reduced space $Rep^{\lambda}_{\overline{Q}}(v,w) := \Phi^{-1}(\mathcal{O}_\lambda)//_G$ induced by the action of the complex reductive group $G = \prod_k GL(V_k)_{k\in Q_0}$. We know from the the results of F. Kirwan [8],R. Sjamaar [22], H. Nakajima [19], and P. Heinzner [6], that the orbit spaces of the reduced space $Rep^{\lambda}_{\overline{Q}}(v,w)$ present some difficulties, since in general they are not closed orbits due to the fact that action of the complex reductive Lie group G is not a proper action. However, it has been shown that because of the noncompactness of its orbits, one could first define the neighborhood of the compact K-orbit $K.m$, and then extend it to the neighborhood of G-orbits $G.m$.

Definition 2.8 *(P. Heinzner [6], K. Fritzche and H. Grauert [3]) A subset U of a G- space $Rep_{\overline{Q}}(v,w)$ is called a saturated subset with respect to the quotient map*

$$\pi_R : Rep_{\overline{Q}}(v,w) \longrightarrow Rep_{\overline{Q}}(v,w)//_G,$$

if $\pi_R^{-1}(\pi_R(U)) = U$, that is, for every $x \in U$, the closure of the orbit $G.x$ is contained in U.

The difficulty encountered in the construction of slices is basically in controlling the behavior of the action of the complex reductive Lie group G on $\Phi^{-1}(\mathcal{O}_\lambda)$ at its "infinity"(see R. Sjamaar in [22]). But R. Sjamaar has shown that for any momentum map there are always slices S such that GS is saturated with respect to the corresponding canonical map.

Definition 2.9 *(E. Lermann, R. Sjamaar [11], Definition 4)*
A slice at a point $x = (\varphi,a,b) \in Rep_{\overline{Q}}(v,w)$ for the G- action is a locally closed subspace S of $Rep_{\overline{Q}}(v,w)$ such that the following conditions hold:

1. *$x \in S$.*

2. *The saturated subspace GS of S is open in S.*

3. *S is invariant under the action of the stabilizer G_x, i.e., $G_xS \subset S$.*

4. *The G- equivariant map from the associated bundle $G \times_{G_x} S$ into $Rep_{\overline{Q}}(v,w)$, that is,*

$$\pi : G \times_{G_x} S \longrightarrow Rep_{\overline{Q}}(v,w)$$
$$[g,y] \mapsto gy \in GS,$$

which sends an equivalence class $[g,y]$ to the point $gy \in GS$ is an isomorphism onto GS.

Another definition needed for the construction of the slices for the complex reductive Lie group G is that of orbital convexity.

Definition 2.10 *A subset U of a G- space $Rep_{\overline{Q}}(v,w)$ is called orbitally convex with respect to the G- action: $\Psi : G \times Rep_{\overline{Q}}(v,w) \longrightarrow Rep_{\overline{Q}}(v,w)$, if U is K- invariant such that for every $u \in U$ and for all $\xi \in \mathcal{K} = LieK$, the intersection of the curve $\{exp(\sqrt{-1}t\xi)u : t \in \mathbb{R}\}$ with U is connected. Equivalently, U is orbitally convex if and only if $\forall u \in U$ and $\forall \xi \in \mathcal{K}$ if both u and $exp(\sqrt{-1}\xi)u$ lie in U, then $exp(\sqrt{-1}t\xi)u \in U$, for all $t \in [0,1]$.*

An important result of P. Heinzner [6] gives conditions for the extension of a K- equivariant map defined on an orbitally convex set to a G-equivariant map, where G is the complexification of a maximal subgroup K. We state this key result (P. Heinzner [6], and R. Sjamaar [22]) for our complex symplectic manifold $(Rep_{\overline{Q}}(v,w), \omega, J)$ with tamed ω-compatible complex structure J.

Theorem 2.11 *Let $(Rep_{\overline{Q}}(v,w), \omega, J)$, $(Rep_{\overline{Q}}(v',w'), \omega', J')$ be complex symplectic manifolds with tamed ω-compatible, ω'-compatible complex structures J and J' respectively, where $v,w,v',w' \in \mathbb{Z}^{Q_0}_{\geq 0}$ are the respective dimension vectors of the representation spaces $(V,\varphi),(W,\psi),(V',\varphi')$ and (W',ψ'). Let G be a reductive complex Lie group with actions*

$$\begin{array}{llll} \Psi: & G \times Rep_{\overline{Q}}(v,w) & \longrightarrow & Rep_{\overline{Q}}(v,w) \\ & (g,(\varphi,a,b)) & \mapsto & g(\varphi,a,b) \\ \text{and} & & & \\ \tilde{\Psi} & G \times Rep_{\overline{Q}}(v',w') & \longrightarrow & Rep_{\overline{Q}}(v',w') \\ & (h,(\varphi,a',b')) & \mapsto & h(\varphi',a',b'). \end{array}$$

Suppose $O \subset Rep_{\overline{Q}}(v,w)$ is an orbitally convex subset and $\rho : O \longrightarrow Rep_{\overline{Q}}(v',w')$ a K-equivariant holomorphic map. Then ρ can be uniquely extended to a G-equivariant holomorphic map $\rho^{\mathbb{C}} : GO \longrightarrow Rep_{\overline{Q}}(v',w')$. Furthermore, if the image $\rho(O)$ of O is open and orbitally convex in $Rep_{\overline{Q}}(v',w')$ and $\rho : O \longrightarrow \rho(O) \subset Rep_{\overline{Q}}(v',w')$ is biholomorphic, then the extension $\rho^{\mathbb{C}} : GO \longrightarrow Rep_{\overline{Q}}(v',w')$ is biholomorphic onto $G\rho(O)$.

Proof. The proof is similar to that in Sjamaar [22]. Here we give only a brief sketch to indicate the flavour of the proof. To extend ρ equivariantly to $\rho^{\mathbb{C}}$, write $\rho^{\mathbb{C}}(kexp(\sqrt{-1}\xi)z) = kexp(\sqrt{-1}\xi)\rho(z)$ for all $z \in O, k \in K$ and $\xi \in \mathcal{K}$. Next we need to verify that it is well-defined. To do this, pick $z \in O$ and $\xi \in \mathcal{K}$ such that $exp(\sqrt{-1}\xi)z \in O$. Then the assumption that $exp(\sqrt{-1}t\xi) \in O$, for $0 \leq t \leq 1$, implies that $\rho(exp(\sqrt{-1}t\xi)z)$ is well-defined for $0 \leq t \leq 1$. Let us now define two curves $\lambda(t)$ and $\eta(t)$ in $Rep_{\overline{Q}}(v',w')$ as follows:

$$\lambda(t) := \rho(exp(\sqrt{-1}t\xi)z)$$

and

$$\eta(t) := exp(\sqrt{-1}t\xi)\rho(z), \text{ for } 0 \leq t \leq 1.$$

Then $\lambda(t)$ and $\eta(t)$ are integral curves for the vector fields $\rho_*(\sqrt{-1}\xi)_{Rep_{\overline{Q}}(v,w)}$ and $(\sqrt{-1}\xi)_{Rep_{\overline{Q}}(v',w')}$ respectively, both having $\rho(z)$ as initial value. Because ρ is K-equivariant we obtain $\rho_*\xi_{Rep_{\overline{Q}}(v,w)} = \xi_{Rep_{\overline{Q}}(v',w')}$. Since ρ is holomorphic we get $\rho_*(\sqrt{-1}\xi)_{Rep_{\overline{Q}}(v,w)} = \rho_*(J\xi_{Rep_{\overline{Q}}(v,w)}) = J\rho_*\xi_{Rep_{\overline{Q}}(v,w)} = J\xi_{Rep_{\overline{Q}}(v',w')}$. Hence $\lambda(t) = \eta(t)$, i.e., $\rho(exp(\sqrt{-1}t\xi)z) = exp(\sqrt{-1}t\xi)\rho(z)$ for $t \in [0,1]$. It follows that $\forall z \in O$ and $\forall \xi \in \mathcal{K}$ such that $exp(\sqrt{-1}\xi)z \in O$, we have $\rho(exp(\sqrt{-1}\xi)z) = exp(\sqrt{-1}\xi)\rho(z)$. One can now without further difficulty deduce that $\rho^{\mathbb{C}}$ is well-defined. Therefore, if the image $\rho(O)$ of O is open and orbitally convex in $Rep_{\overline{Q}}(v',w')$ so that $\rho : O \longrightarrow \rho(O) \subset Rep_{\overline{Q}}(v',w')$ is biholomorphic then the inverse ρ^{-1} has a biholomorphic extension $(\rho^{-1})^{\mathbb{C}} : G\rho(O) \longrightarrow GO$, and by uniqueness it must be the inverse of the map $\rho^{\mathbb{C}} : G\rho(O) \longrightarrow Rep_{\overline{Q}}(v',w')$. $\Diamond$

It is important to point out here that the orbital convexity of $O \subset Rep_{\overline{Q}}(v,w)$ is crucial in the above theorem. Without it, it is not necessarily true that $\rho(exp(\sqrt{-1}t\xi)z)$ is equal to $exp(\sqrt{-1}t\xi)\rho(z)$ for all t for which $exp(\sqrt{-1}t\xi)z \in O$. If we consider only the connected component $\mathbb{I}_0$ of the set $\mathbb{I} = \{t \in \mathbb{R} : exp(\sqrt{-1}t\xi)z \in O\}$ then $\forall t \in \mathbb{I}_0$, we obtain:

$$\rho(exp(\sqrt{-1}t\xi)z) = exp(\sqrt{-1}t\xi)\rho(z).$$

In order to get a good understanding of the orbit structure of the complex reductive Lie group $G := \prod_{k\in Q_0} GL_{\mathbb{C}}(V_k)$ on $Rep_{\overline{Q}}(v,w)$, it is important to be able to construct slices for the G-action at points that lie in the λ-level set of the momentum map

$$\Phi : Rep_{\overline{Q}}(v,w) \longrightarrow \mathcal{G}^*$$

for $\lambda \in O_\lambda \subset \mathcal{G}^*$ a co-adjoint orbit of the co-adjoint action
$Ad^* : G \times \mathcal{G}^* \longrightarrow \mathcal{G}^*$ on the dual Lie algebra $\mathcal{G}^*$ of the Lie algebra $\mathcal{G} = Lie(G)$ of G. The result which makes this possible is the following general theorem of R. Sjamaar [22], the Holomorphic Slice Theorem. We will state the theorem in the general form as proved by R. Sjamaar [22], but our applications will be to $(Rep_{\overline{Q}}(v,w), \omega, J)$ a complex symplectic manifold with ω-compatible tamed complex structure J, which is at the same time a Kähler and hyper-Kähler manifold. We introduce the Kähler quotient of $Rep_{\overline{Q}}(v,w)$ by G, following the construction introduced by F. Kirwan [8]. There is also the stratified symplectic structure of the symplectic quotient introduced by R. Sjamaar and E. Lerman [23]. The Holomorphic Slice Theorem should be a useful tool to help us understand some relationships between these constructions of quotients. The fact that G is a non-compact group and thus never acts properly on $Rep_{\overline{Q}}(v,w)$ coupled with the fact that the orbits are not closed raises serious difficulties. The result of R. Sjamaar we have been alluding to is the following theorem.

Theorem 2.12 *(Holomorphic Slice Theorem: R. Sjamaar [22]) Let X be a Kähler manifold and G a complex reductive Lie group. Let K be a maximal compact real Lie subgroup of G such that G is the complexification of K. Suppose the action*

$$\begin{array}{rccc} \Psi : & K \times X & \longrightarrow & X \\ & (k,x) & \mapsto & \Psi(k,x) := kx, \end{array}$$

is Hamiltonian and x is any point in X such that the K-orbit through x is isotropic. Then there exists a Slice at x for the G-action:

$$\begin{array}{rccc} \Psi^{\mathbb{C}} : & G \times X & \longrightarrow & X \\ & (g,x) & \mapsto & \Psi^{\mathbb{C}}(g,x) := gx. \end{array}$$

Proof. See R. Sjamaar [22]. $\diamondsuit$

In our case if $X := Rep_{\overline{Q}}(v,w)$ and S is a slice at some point $m = (\varphi, a, b) \in Rep_{\overline{Q}}(v,w)$ then kS is a slice at km for any $k \in K$ so the theorem should give us a slice at any point $m = (\varphi, a, b) \in Rep_{\overline{Q}}(v,w)$ such that the G-orbit through m contains an isotropic K-orbit. Furthermore, if $\tilde{K}$ is another maximal compact real Lie subgroup of G, we immediately see that $\tilde{K}$ is conjugate to K by $g \in G$, i.e., $\tilde{K} = gKg^{-1}$ and so $\tilde{K}$ preserves the symplectic form and we get a K-momentum map $\tilde{\Phi} = (Ad_g^*) \circ \Phi \circ g^{-1}$ with $\Phi : X \longrightarrow \mathcal{K}^*$ an Ad^*-equivariant momentum map for the K-action. This means essentially that the statement of the theorem is independent of the choice of the compact maximal real subgroup K of G. We will now explain the shifting trick due to V. Guillemin and S. Sternberg (see R. Sjamaar and E. Lerman [23]). Let (M, ω) be a symplectic manifold and K a compact Lie group such that the K-action $\Psi : K \times M \longrightarrow M \ : \ (k,m) \mapsto \Phi(k,m) = km$ is proper and Hamiltonian with associated momentum map $\Phi : M \longrightarrow \mathcal{K}^*$, where $\mathcal{K}^*$ is the dual Lie algebra of the Lie algebra $\mathcal{K} = \mathrm{Lie}(K)$ of K. For any point $\lambda \in \mathcal{K}^*$, let $\mathcal{O}_\lambda$ denote the co-adjoint orbit through λ. If λ is a regular value of the momentum map the pre-image $\Phi^{-1}(\mathcal{O}_\lambda)$ of the co-adjoint orbit is a K-invariant submanifold of M. Since the image of the linear map $T_q\Phi : T_qM \longrightarrow T_{\Phi(q)}\mathcal{K}^* = \mathcal{K}^*$, for all $q \in M$ is equal to the annihilator $Ann\,\mathcal{K}_q$, i.e., $ImT_q\Phi = Ann\,\mathcal{K}_q$, it follows that, the action of K on $\Phi^{-1}(\mathcal{O}_\lambda)$, $\Psi : K \times \Phi^{-1}(\mathcal{O}_\lambda) \longrightarrow \Phi^{-1}(\mathcal{O}_\lambda)$ is locally free, that is, all stabilizers are discrete. J. Marsden and A. Weinstein [16], defined the reduced space M_λ to be the quotient orbifold:

$$M_\lambda = \Phi^{-1}(\mathcal{O}_\lambda)//_K.$$

They proved the following fact about M_λ:
There is a unique symplectic form ω_λ on the space M_λ such that the pull-back of ω_λ to $\Phi^{-1}(\mathcal{O}_\lambda)$ is the same as the restriction of the form ω to $\Phi^{-1}(\mathcal{O}_\lambda)$. In the case where λ is not a regular value of Φ the set $\Phi^{-1}(\mathcal{O}_\lambda)$ is not in general a manifold. Thus from the fact $ImT_q\Phi = Ann\,\mathcal{K}_q$ we deduce that the action of K on $\Phi^{-1}(\mathcal{O}_\lambda)$ is not locally free, however, the definition of $M_\lambda = \Phi^{-1}(\mathcal{O}_\lambda)//_K$ still makes sense topologically. Consider, the case $\lambda \neq 0$ is a regular value of $\Phi : M \longrightarrow \mathcal{K}^*$ and introduce the symplectic manifold $M \times \mathcal{O}_\lambda^-$, the symplectic product of M with the co-adjoint orbit $\mathcal{O}_\lambda$ through λ endowed with the opposite Kostant-Kirillov-Souriau symplectic form. The diagonal action of K on $M \times \mathcal{O}_\lambda^-$ is Hamiltonian with momentum map Φ_λ given by $\Phi_\lambda(m,\lambda) := \Phi(m) - \lambda$. It is not difficult to see that the cartesian projection $Proj : M \times \mathcal{O}_\lambda^- \longrightarrow M$ restricts to an equivariant bijection $\Phi_\lambda^{-1}(0) \xrightarrow{\simeq} \Phi_\lambda^{-1}(\mathcal{O}_\lambda)$. Hence $Proj : M \times \mathcal{O}_\lambda^- \longrightarrow M$ descends to a bijection between reduced spaces $\underline{Proj} : (M \times \mathcal{O}_\lambda^-)_0 \xrightarrow{\simeq} M_\lambda$ which is an isomorphism of the reduced spaces. The idea of the proof of the isomorphism is based on showing the isomorphism of Poisson algebras:

$$\underline{Proj}^* : C^\infty(M_\lambda, \mathbb{R}) \longrightarrow C^\infty((M_\lambda \times \mathcal{O}_\lambda^-)_0, \mathbb{R}).$$

It is then easily seen that the smooth functions on $(M_\lambda \times \mathcal{O}_\lambda^-)_0$ pull-back to smooth functions on M_λ thus establishing the injectivity. The surjectivity proof is slightly involved, we will skip the details here. We need the following consequence of the Holomorphic Slice Theorem (see (R. Sjamaar [22])).

Theorem 2.13 *Let $m = (\varphi, a, b) \in Rep_{\overline{Q}}(v,w)$ be any fixed point of the G-action*

$$\Psi : G \times Rep_{\overline{Q}}(v,w) \longrightarrow Rep_{\overline{Q}}(v,w).$$

Then G can be linearized in a neighborhood of m such that there exists a G-invariant open neighborhood U of m in $Rep_{\overline{Q}}(v,w)$, a G-invariant neighborhood $\tilde{U}$ of the origin 0 in the tangent space $T_m Rep_{\overline{Q}}(v,w)$ and a biholomorphic G-equivariant map $\rho : U \longrightarrow \tilde{U}$.

Proof. (See R. Sjamaar [22] for the general case). Note that a fixed point of the G-action is clearly isotropic and apply the Holomorphic Slice Theorem. ◇

Assuming that the G-orbits are isotropic orbits, and the momentum map $\Phi : Rep_{\overline{Q}}(v,w) \longrightarrow \mathcal{G}^*$ is admissible, i.e., if the flow F_t is defined $\forall t \geq 0$. F. Kirwan has shown that the set of stable points $Rep_{\overline{Q}}(v,w)^s$ has the property that the path $F_t(m)$ for any $m = (\varphi, a, b) \in Rep_{\overline{Q}}(v,w)^s$ has a limit point in $\Phi^{-1}(0)$ so that for all $m = (\varphi, a, b) \in Rep_{\overline{Q}}(v,w)$ the limit $F_\infty(m) = \lim_{t \to \infty} F_t(m)$ exists and the restriction map: $F_\infty|_{Rep_{\overline{Q}}(v,w)^s} \longrightarrow \Phi^{-1}(0)$ is surjective. H. Nakajima [18] has shown that the set of stable points and the set of the semi-stable points lie in the same space. And for any point $m = (\varphi, a, b) \in Rep_{\overline{Q}}(v,w)$, its neighborhood U is saturated. We can now state a result of F. Kirwan [8] concerning the orbit structure of the semistable set as adapted to our particular situation. We follow the statement in R. Sjamaar [22], Proposition 2.2.

Theorem 2.14 *Let $Rep_{\overline{Q}}(v,w)^{ss} \subset Rep_{\overline{Q}}(v,w)$. In the statements that follow, we observe the following conventions: "closed" will mean closed in the relative topology on $Rep_{\overline{Q}}(v,w)^{ss}$ and "closure" to mean closure in $Rep_{\overline{Q}}(v,w)^{ss}$.*

1. *The semistable set $Rep_{\overline{Q}}(v,w)^{ss} \subset Rep_{\overline{Q}}(v,w)$ is the smallest G-invariant open subset of $Rep_{\overline{Q}}(v,w)$ containing $\Phi^{-1}(O_\lambda)$ with the complement*

$$Rep_{\overline{Q}}(v,w) \backslash Rep_{\overline{Q}}(v,w)^{ss}$$

a complex analytic subset.

2. *A G-orbit in $Rep_{\overline{Q}}(v,w)$ is closed if and only if it intersects $\Phi^{-1}(O_\lambda)$.*

3. *The intersection of a closed G-orbit with $\Phi^{-1}(O_\lambda)$ consists of precisely one K-orbit.*

4. *For every semistable point $m = (\varphi, a, b) \in Rep_{\overline{Q}}(v,w)^{ss}$, the set $F_\infty(G \cdot m) \subset \Phi^{-1}(O_\lambda)$ consists of precisely one $K-$orbit.*

5. *For any pair of points $x, y \in \Phi^{-1}(O_\lambda)$, which do not lie on the same G-orbit, there exist disjoint G-invariant open subsets U and V of $Rep_{\overline{Q}}(v,w)$ with $x \in U$ and $y \in V$.*

6. *The closure of every G-orbit in $Rep_{\overline{Q}}(v,w)^{ss}$ contains exactly one closed G-orbit.*

Proof. See R. Sjamaar [22] for the general proof which is easily adapted to our particular case.◇

Let $O_{Rep^{\lambda}_{\overline{Q}}(v,w)}$ be the sheaf of holomorphic functions on $Rep^{\lambda}_{\overline{Q}}(v,w)$. We can state a result of R. Sjamaar [22] adapted in our situation.

Theorem 2.15 *The ringed space* $(Rep^{\lambda}_{\overline{Q}}(v,w), O_{Rep^{\lambda}_{\overline{Q}}(v,w)})$ *is a complex analytic space.*

Proof. The proof apart from minor adaptation follows exactly the line of proof of that of theorem 2.6 in R. Sjamaar [22]. Therefore, we omit the easy changes.◊

With the above definitions and the R. Sjamaar Holomorphic Slice Theorem in place, the reduced space $Rep^{\lambda}_{\overline{Q}}(v,w)$ has a separation property, closed orbits, and it is also stable. Consider the orbit $[(\varphi,a,b)] \in Rep^{\lambda}_{\overline{Q}}(v,w)$ with the point $(\varphi,a,b) \in \Phi^{-1}(O_\lambda)$ as its representative. Let H be the stabilizer of (φ,a,b) in the compact real Lie group K, and define $H^{\mathbb{C}} = H \otimes_{\mathbb{R}} \mathbb{C}$ its complexification in the complex reductive Lie group G. Then, following the results of R. Sjamaar [22] the stratification of $Rep^{\lambda}_{\overline{Q}}(v,w)$ by K-orbit types induces its stratification by G-orbit types. Hence, each stratum

$$Rep^{\lambda}_{\overline{Q}}(v,w)_{(H^{\mathbb{C}})} := (\Phi^{-1}(O_\lambda) \cap Rep_{\overline{Q}}(v,w)_{(H^{\mathbb{C}})})//G,$$

is a complex manifold and its closure is a complex -analytic subvariety of $Rep^{\lambda}_{\overline{Q}}(v,w)$, where

$$Rep_{\overline{Q}}(v,w)_{(H^{\mathbb{C}})} := \{(\varphi,a,b) \in Rep_{\overline{Q}}(v,w) : \quad gG_{(\varphi,a,b)}g^{-1} = H^{\mathbb{C}}, \quad \forall g \in G\},$$

is the manifold of points of orbit types $(H^{\mathbb{C}})$, and $G_{(\varphi,a,b)}$ is the stabilizer of G at the point $(\varphi,a,b) \in \Phi^{-1}(O_\lambda)$. To proof this claim, see the proof of the theorem 2.9 in R. Sjamaar [22].

2.1 Smooth Structures on the Reduced Spaces

Let

$$\mathcal{F}(Rep_{\overline{Q}}(v,w)) := C^\infty(Rep_{\overline{Q}}(v,w),\mathbb{R}),$$

be the algebra of smooth functions on $Rep_{\overline{Q}}(v,w)$. The extension of scalars gives $C^\infty(Rep_{\overline{Q}}(v,w),\mathbb{C}) := C^\infty(Rep_{\overline{Q}}(v,w),\mathbb{R}\otimes_{\mathbb{R}}\mathbb{C})$, the algebra of smooth complex valued functions on $Rep_{\overline{Q}}(v,w)$.

Let $C^\infty(Rep_{\overline{Q}}(v,w),\mathbb{R})^G$ and $C^\infty(Rep_{\overline{Q}}(v,w),\mathbb{C})^G$ be the spaces of smooth G-invariant functions on $Rep_{\overline{Q}}(v,w)$, which are both Poisson subalgebras of the Poisson algebras of $C^\infty(Rep_{\overline{Q}}(v,w),\mathbb{R})$ and $C^\infty(Rep_{\overline{Q}}(v,w),\mathbb{C})$ respectively see [23], [2], and [24]. By the injective morphism induced by complexification we have the map:

$$(C^\infty(Rep_{\overline{Q}}(v,w),\mathbb{R})^G,\{.,.\},.) \xrightarrow{\delta} (C^\infty(Rep_{\overline{Q}}(v,w),\mathbb{C})^G,\{.,.\},.)$$
$$f^{\mathbb{R}} \mapsto f^{\mathbb{C}}$$

Then for every $F,H \in C^\infty(Rep_{\overline{Q}}(v,w),\mathbb{R})^G$,

$$\Psi^*_g\{F,H\} = \{\Psi^*_g F, \Psi^*_g H\} = \{F,H\},$$

since F,H are G-invariant. Hence $\{F,H\} \in \mathcal{F}(Rep_{\overline{Q}}(v,w))^G := C^\infty(Rep_{\overline{Q}}(v,w),\mathbb{R})^G$. It is not difficult to see that

$$\Psi^*_g(F\cdot H) = \Psi^*_g F \cdot \Psi^*_g H = F\cdot H.$$

This shows that $F \cdot H \in \mathcal{F}(Rep_{\overline{Q}}(v,w)) = C^{\infty}(Rep_{\overline{Q}}(v,w),\mathbb{R})^G$ and thus $(C^{\infty}(Rep_{\overline{Q}}(v,w),\mathbb{R})^G, \{\cdot,\cdot\},\cdot)$ is a Poisson subalgebra of the Poisson algebra $(C^{\infty}(Rep_{\overline{Q}}(v,w),\mathbb{R}), \{\cdot,\cdot\},\cdot)$.
Now let O_λ be the coadjoint orbit through the singular value $\lambda \in \mathcal{G}^*$ of the momentum map $\Phi : Rep_{\overline{Q}}(v,w) \longrightarrow \mathcal{G}^*$. Let

$$\pi_R : Rep_{\overline{Q}}(v,w) \longrightarrow Rep_{\overline{Q}}(v,w)//G$$

be the quotient map of the G-action:

$$\begin{aligned} \Psi : \quad & G \times Rep_{\overline{Q}}(v,w) \longrightarrow Rep_{\overline{Q}}(v,w) \\ & (g,(\varphi,a,b)) \mapsto \Psi(g,(\varphi,a,b)) = (g\varphi g^{-1}, ga, bg^{-1}). \end{aligned}$$

The affine algebraic variety $\Phi^{-1}(O_\lambda) \subset Rep_{\overline{Q}}(v,w)$ induces the G-action

$$\Psi|_{\Phi^{-1}(O_\lambda)} : G \times \Phi^{-1}(O_\lambda) \longrightarrow \Phi^{-1}(O_\lambda).$$

Then the quiver variety $Rep^{\lambda}_{\overline{Q}}(v,w) := \Phi^{-1}(O_\lambda)//_G = \pi_\Phi(\Phi^{-1}(O_\lambda))$, is a stratified singular quiver variety and is stratified by orbit type, see [1], [2], [23], [17], where $\pi_\Phi : \Phi^{-1}(O_\lambda) \longrightarrow \Phi^{-1}(O_\lambda)//_G$ is a projection map. The smooth functions on $Rep^{\lambda}_{\overline{Q}}(v,w)$ are smooth complex-valued functions if the following diagram commutes:

$$\begin{array}{ccc} Rep_{\overline{Q}}(v,w) & \overset{\pi_R}{\longrightarrow} & Rep^{\lambda}_{\overline{Q}}(v,w) \\ & {\scriptstyle f}\searrow \quad \swarrow{\scriptstyle f_\lambda} & \\ & \mathbb{C}, & \end{array}$$

$$i.e., \pi_R^* f_\lambda = f|_{\Phi^{-1}(O_\lambda)}.$$

Where f is a G-invariant function. It then follows that

$$C^{\infty}(Rep^{\lambda}_{\overline{Q}}(v,w),\mathbb{C}) := C^{\infty}(Rep_{\overline{Q}}(v,w),\mathbb{C})^G / I(\Phi^{-1}(O_\lambda)),$$

where $I(\Phi^{-1}(O_\lambda)) := \{f \in C^{\infty}(Rep_{\overline{Q}}(v,w),\mathbb{C})^G : f|_{\Phi^{-1}O_\lambda} = 0\}$ is the ideal of smooth G-invariant complex-valued functions which vanish identically on $\Phi^{-1}(O_\lambda)$. Since the requirement of orbital convexity and isotropic orbit force our orbits to be closed orbits, then the orbits O_λ are closed. Thus $\Phi^{-1}(O_\lambda)$ is a closed affine variety of $Rep_{\overline{Q}}(v,w)$, and $I(\Phi^{-1}(O_\lambda))$ is a closed subspace of $C^{\infty}(Rep_{\overline{Q}}(v,w),\mathbb{C})$. The ideal $I(\Phi^{-1}(O_\lambda))$ is a Poisson ideal of the Poisson Lie algebra $(C^{\infty}(Rep_{\overline{Q}}(v,w),\mathbb{C}),\{.,.\},.)$. We thus have the following Lemma.

Lemma 2.16 *Let $(C^{\infty}(Rep_{\overline{Q}}(v,w)^G,\mathbb{C}),\{.,.\},.)$ be a Poisson subalgebra of the Poisson algebra $(C^{\infty}(Rep_{\overline{Q}}(v,w),\mathbb{C}),\{.,.\},.)$ and suppose that for every*

$$H \in C^{\infty}(Rep_{\overline{Q}}(v,w),\mathbb{C})^G$$

the flow of the Hamiltonian vector field X_H corresponding to the G-invariant function H on $Rep_{\overline{Q}}(v,w)$ preserves $\Phi^{-1}(O_\lambda)$. Then the ideal $I(\Phi^{-1}(O_\lambda)) := \{f \in C^{\infty}(Rep_{\overline{Q}}(v,w),\mathbb{C})^G : f|_{\Phi^{-1}(O_\lambda)} = 0\}$ is a Poisson ideal of $C^{\infty}(Rep_{\overline{Q}}(v,w),\mathbb{C})^G$.

Proof. We give only a sketch of the proof as the details are easy to see. Let $F \in (C^\infty(Rep_{\overline{Q}}(v,w)^G, \mathbb{C}), \{.,.\},.)$, and $m = (\varphi, a, b) \in \Phi^{-1}(O_\lambda)$ and $H \in I(\Phi^{-1}(O_\lambda))$. Let $\gamma : \mathbb{R} \longrightarrow \Phi^{-1}(O_\lambda)$ be an integral curve of X_F the Hamiltonian vector field of F with $\gamma(0) = m$. Then $\gamma(t) \in \Phi^{-1}(O_\lambda)$ for all $t \in I_\varepsilon := (-\varepsilon, \varepsilon) \subset \mathbb{R}$, $\varepsilon > 0$, containing $0 \in \mathbb{R}$. Therefore, $H(\gamma(t)) = 0$, for all $t \in I_\varepsilon$ and

$$0 = \frac{d}{dt}|_{t=0} H(\gamma(t)) = \{F, H\}(m).$$

This shows that $\{F, H\} \in I(\Phi^{-1} O_\lambda)$. $\Diamond$

Using the quotient map

$$\pi_R : Rep_{\overline{Q}}(v,w) \longrightarrow Rep_{\overline{Q}}(v,w)//G,$$

we define a Poisson bracket on $C^\infty(Rep^{\lambda}_{\overline{Q}}(v,w), \mathbb{C})$ by

$$\pi_R^*\{F_\lambda, H_\lambda\}_\lambda := \{F, H\}|\Phi^{-1}(O_\lambda)$$

such that

$$\pi_R^* F_\lambda = F|\Phi^{-1}(O_\lambda) \text{ and } \pi_R^* H_\lambda = H|\Phi^{-1}(O_\lambda).$$

R. Sjamaar and E. Lermann in [23], and Sniatyck in [24] show that the Poisson structure $\{.,.\}_\lambda$ makes sense, we adapt their Lemma in our situation.

Lemma 2.17 *Let $Rep^{\lambda}_{\overline{Q}}(v,w)$ be our quiver variety. The Poisson bracket*

$$\{.,.\}_\lambda : C^\infty(Rep^{\lambda}_{\overline{Q}}(v,w)) \times C^\infty(Rep^{\lambda}_{\overline{Q}}(v,w)) \longrightarrow C^\infty(Rep^{\lambda}_{\overline{Q}}(v,w)),$$

defined above on $C^\infty(Rep^{\lambda}_{\overline{Q}}(v,w))$ is:

1. *A well-defined Poisson bracket and*

2. *Nondegenerate, i.e., its only Casimir functions are locally constant.*

Proof. We only give here, an idea of the proof of this Lemma. Complete details of the proof are analogous to the ones in the papers [23], [24].
Firstly, we let $H, \tilde{H} \in C^\infty(Rep_{\overline{Q}}(v,w), \mathbb{C})^G$ such that $\pi_R^* H_\lambda = \tilde{H}|\Phi^{-1}(O_\lambda)$, for any $H_\lambda \in C^\infty(Rep^{\lambda}_{\overline{Q}}(v,w))$. Then we show that $\{F, H - \tilde{H}\} \in \Phi^{-1}(O_\lambda)$.
Secondly, to show that $\{.,.\}_\lambda$ is a nondegenerate it suffices to show that if $H \in C^\infty(Rep_{\overline{Q}}(v,w), \mathbb{C})^G$ such that $\{F, H\}(\varphi, a, b) = 0$, for any point $(\varphi, a, b) \in \Phi^{-1}(O_\lambda)$, and $F \in C^\infty(Rep_{\overline{Q}}(v,w), \mathbb{C})^G$, then $H|\Phi^{-1}(O_\lambda)$ is locally constant. $\Diamond$

To study the dynamics of the reduced space, we follow L. Bates and E. Lerman [1]in defining the general flows and Hamiltonian flows on the reduced space $Rep^{\lambda}_{\overline{Q}}(v,w)$.

Definition 2.18 *([1])*

1. *A smooth curve in the quiver variety* $Rep^{\lambda}_{\overline{Q}}(v,w)$ *is a continuous map*

$$\gamma : I_{\varepsilon} := \{-\varepsilon, \varepsilon\} \subset \mathbb{R} \longrightarrow Rep^{\lambda}_{\overline{Q}}(v,w),$$

with the property that given any smooth function $F_{\lambda} \in (C^{\infty}(Rep^{\lambda}_{\overline{Q}})(v,w), \mathbb{C})$ *the composite* $F_{\lambda} \circ \gamma(t)$ *is a smooth function on* $I_{\varepsilon} = (-\varepsilon, \varepsilon) \subset \mathbb{R}$.

2. *A smooth flow* $\{\varphi_s\}$ *is similarly defined.*

$$\varphi_s : Rep^{\lambda}_{\overline{Q}}(v,w) \longrightarrow Rep^{\lambda}_{\overline{Q}}(v,w)$$

is a one-parameter group of homeomorphisms such that for any $F_{\lambda} \in C^{\infty}(Rep^{\lambda}_{\overline{Q}}(v,w), \mathbb{C})$ *and any parameter s, the composite morphism* $F_{\lambda} \circ \varphi_s \in C^{\infty}(Rep_{\overline{Q}}(v,w), \mathbb{C})$ *and for each point* $m = [(\varphi, a, b)] \in Rep^{\lambda}_{\overline{Q}}(v,w)$ *the curve* $s \mapsto \varphi_s(m)$ *is a smooth curve.*

3. *A Hamiltonian flow of a smooth function* $F_{\lambda} \in C^{\infty}(Rep^{\lambda}_{\overline{Q}}(v,w), \mathbb{C})$ *is a smooth map*

$$\varphi : \mathbb{R} \times Rep^{\lambda}_{\overline{Q}}(v,w) \longrightarrow Rep^{\lambda}_{\overline{Q}}(v,w),$$

where $(\mathbb{R}, +)$ *is the additive group acting on* $Rep^{\lambda}_{\overline{Q}}(v,w)$ *such that for any point* $m = [(\varphi, a, b)] \in Rep^{\lambda}_{\overline{Q}}(v,w)$ *and any smooth function* $H_{\lambda} \in C^{\infty}(Rep^{\lambda}_{\overline{Q}}(v,w), \mathbb{C})$, *we have:*

$$\frac{dH_{\lambda}(\varphi_s(m))}{ds} = \{F_{\lambda}, H_{\lambda}\}_{\lambda}(\varphi_s(m)),$$

where $\{\cdot,\cdot\}_{\lambda}$ *is the Poisson bracket on the algebra* $C^{\infty}(Rep^{\lambda}_{\overline{Q}}(v,w), \mathbb{C})$.

Note that since the quiver variety $Rep^{\lambda}_{\overline{Q}}(v,w)$ is not a smooth variety, the equation:

$$\frac{dH_{\lambda}(\varphi_s(m))}{ds} = \{F_{\lambda}, H_{\lambda}\}_{\lambda}(\varphi_s(m))$$

is not in general a system of ordinary differential equations in a coordinate-free representation and so one expects non-uniqueness of solutions on any non-Hausdorff variety.
The definitions above together with Lemma 5 [1], allow us to deduce that not only the singular variety $Rep^{\lambda}_{\overline{Q}}(v,w)$ has a separation property, but that the Hamiltonian flows are unique.

3 Singular Reduction Theorem

In this section, we state and prove the main result of this article, that is the singular reduction theorem. Let G be the complex reductive Lie group with K its maximal compact real Lie subgroup .

Theorem 3.1 (Singular Reduction Theorem) *Let* $G = \sqcap_{k \in Q_0} GL_{\mathbb{C}}(V_k)$ *be a complex reductive Lie group and K a maximal compact real Lie subgroup of G. Let*

$$\Psi : K \times Rep_{\overline{Q}}(v,w) \longrightarrow Rep_{\overline{Q}}(v,w)$$

be a proper and Hamiltonian action on the symplectic manifold $(Rep_{\overline{Q}}(v,w), \omega)$ *which admits a coadjoint equivariant momentum map*

$$\Phi : Rep_{\overline{Q}}(v,w) \longrightarrow \mathcal{K}^*.$$

Furthermore, suppose that the coadjoint orbit $\mathcal{O}_{\lambda_{\mathcal{K}^*}}$ *through the singular value* $\lambda_{\mathcal{K}^*} \in \mathcal{K}^*$ *is locally closed and let the extension*

$$\Psi^{\mathbb{C}} : K^{\mathbb{C}} \times Rep_{\overline{Q}}(v,w) \longrightarrow Rep_{\overline{Q}}(v,w)$$

be a symplectic action of G, so that for any point $q = (\varphi, a, b) \in Rep_{\overline{Q}}(v,w)$ *the K-orbit through q is isotropic. Then on the singular reduced space* $Rep_{\overline{Q}}^{\lambda_{\mathcal{K}^*}}(v,w) := \Phi^{-1}(\mathcal{O}_{\lambda_{\mathcal{K}^*}})//K$, *there is a nondegenerate Poisson algebra*

$$(C^\infty(Rep_{\overline{Q}}^{\lambda_{\mathcal{K}^*}}(v,w), \mathbb{C}), \{\cdot,\cdot\}_{\lambda_{\mathcal{K}^*}}, \cdot),$$

and $Rep_{\overline{Q}}^{\lambda_{\mathcal{K}^*}}(v,w)$ *is a locally finite union of symplectic manifolds called symplectic pieces or strata.*
The flow of a Hamiltonian derivation corresponding to a smooth function on $Rep_{\overline{Q}}^{\lambda_{\mathcal{K}^*}}(v,w)$ *preserves the decomposition of*

$$Rep_{\overline{Q}}^{\lambda_{\mathcal{K}^*}}(v,w) := \blacksquare_{(H)} Rep_{\overline{Q}}^{\lambda_{\mathcal{K}^*}}(v,w)_{(H)}$$

into symplectic pieces or strata and the inclusion of the symplectic pieces or strata: $\iota_{R^{\lambda_{\mathcal{K}^*}}} : Rep_{\overline{Q}}^{\lambda_{\mathcal{K}^*}}(v,w)_{(H)} \longrightarrow Rep_{\overline{Q}}^{\lambda_{\mathcal{K}^*}}(v,w)$ *is a Poison map.*

Proof. The proof of this theorem is made up of the Lemmas 2.16, 2.17, and Theorem 2.12. Much of the proof is derived from similar results in the references (see also R. Sjamaar [22], E. Lerman and R. Sjamaar [23], [11], L. Bates and E. Lerman [1]). We need now only show the following:
(α) the flow of a Hamiltonian derivation corresponding to a smooth function on $Rep_{\overline{Q}}^{\lambda_{\mathcal{K}^*}}(v,w)$ preserves the decomposition of

$$Rep_{\overline{Q}}^{\lambda_{\mathcal{K}^*}}(v,w) = \blacksquare_{(H)} Rep_{\overline{Q}}^{\lambda_{\mathcal{K}^*}}(v,w)_{(H)}$$

into symplectic pieces or strata and
(β) the inclusion of the symplectic pieces or strata

$$\iota_{R^{\lambda_{\mathcal{K}^*}}} : Rep_{\overline{Q}}^{\lambda_{\mathcal{K}^*}}(v,w)_{(H)} \longrightarrow Rep_{\overline{Q}}^{\lambda_{\mathcal{K}^*}}(v,w)$$

is a Poisson map.

To prove (α): Let $f_{\lambda_{\mathcal{K}^*}} \in C^\infty(Rep_{\overline{Q}}^{\lambda_{\mathcal{K}^*}}(v,w)_{(H)}, \mathbb{C})$ be a C^∞-complex valued function on the stratum $Rep_{\overline{Q}}^{\lambda_{\mathcal{K}^*}}(v,w)_{(H)}$ of $Rep_{\overline{Q}}^{\lambda_{\mathcal{K}^*}}(v,w)$ and $\pi_R : Rep_{\overline{Q}}(v,w) \longrightarrow Rep_{\overline{Q}}(v,w)//K$, with

$$\begin{aligned} Rep_{\overline{Q}}^{\lambda_{\mathcal{K}^\star}}(v,w)_{(H)} &:= \pi_R(Rep_{\overline{Q}}(v,w)_{(H)}) \cap \Phi^{-1}(\mathcal{O}_{\lambda_{\mathcal{K}^\star}}) \\ &= \pi_R(\Phi^{-1}(\mathcal{O}_{\lambda_{\mathcal{K}^\star}})_{(H)}), \end{aligned}$$

where $\Phi^{-1}(\mathcal{O}_{\lambda_{\mathcal{K}^\star}})_{(H)} := Rep_{\overline{Q}}(v,w)_{(H)} \cap \Phi^{-1}(\mathcal{O}_{\lambda_{\mathcal{K}^\star}})$.
Then $\pi_R^* f_{\lambda_{\mathcal{K}^\star}} = f|\Phi^{-1}(\mathcal{O}_{\lambda_{\mathcal{K}^\star}})_{(H)}$, for $f \in C^\infty(Rep_{\overline{Q}}(v,w),\mathbb{C})^K$ a smooth K-invariant function on $Rep_{\overline{Q}}(v,w)$. Let $\varphi_t^{X_f}$ be the flow of the vector field X_f. We need to show:

$$\varphi_t^{X_f}(\Phi^{-1}(\mathcal{O}_{\lambda_{\mathcal{K}^\star}})_{(H)}) \subseteq \Phi^{-1}(\mathcal{O}_{\lambda_{\mathcal{K}^\star}})_{(H)}. \tag{3.1}$$

Because

$$\Phi^{-1}(\mathcal{O}_{\lambda_{\mathcal{K}^\star}})_{(H)} := K \cdot (\Phi^{-1}(\mathcal{O}_{\lambda_{\mathcal{K}^\star}})_{(H)} \cap Rep_{\overline{Q}}(v,w)_H),$$

it is enough to show that:

1. $\varphi_t^{X_f}(\Phi^{-1}(\mathcal{O}_{\lambda_{\mathcal{K}^\star}})) \subseteq \Phi^{-1}(\mathcal{O}_{\lambda_{\mathcal{K}^\star}})$
2. $\varphi_t^{X_f}(Rep_{\overline{Q}}(v,w)_H) \subseteq Rep_{\overline{Q}}(v,w)_H$.

The proof of (1) follows since $f \in C^\infty(Rep_{\overline{Q}}(v,w),\mathbb{C})^K$. To see (2) pick any point $q = (\varphi,a,b) \in Rep_{\overline{Q}}(v,w)$. Since $f \in C^\infty(Rep_{\overline{Q}}(v,w),\mathbb{C})^K$, the flow $\varphi_t^{X_f}$ of X_f and the K-action $\Psi_k : Rep_{\overline{Q}}(v,w) \longrightarrow Rep_{\overline{Q}}(v,w)$ commute $\forall k \in K$. Therefore, for every $h \in H$

$$\Psi_h(\varphi_t^{X_f}(q)) = \varphi_t^{X_f}(\Psi_h(q)) = \varphi_t^{X_f}(q) \tag{3.2}$$

This means $\varphi_t^{X_f}(q) \in Rep_{\overline{Q}}(v,w)_{[H]}$. But $Rep_{\overline{Q}}(v,w)_H \subseteq Rep_{\overline{Q}}(v,w)_{[H]}$ is relatively open in $Rep_{\overline{Q}}(v,w)_{[H]}$, so there exists an open neighborhood U of q in $Rep_{\overline{Q}}(v,w)$ with $Rep_{\overline{Q}}(v,w)_H \cap U = Rep_{\overline{Q}}(v,w)_{[H]} \cap U$, i.e., $\varphi_t^{X_f}(Rep_{\overline{Q}}(v,w)_H \cap U) \subseteq Rep_{\overline{Q}}(v,w)_{[H]} \cap U$. The flow $\varphi_t^{X_f}$ induces a homeomorphism of $Rep_{\overline{Q}}(v,w)_H$ in the relative topology, since $\varphi_t^{X_f}$ is a homeomorphism of $Rep_{\overline{Q}}(v,w)$. This implies that $\varphi_t^{X_f}(Rep_{\overline{Q}}(v,w)_H \cap U)$ is a relative open subset of $Rep_{\overline{Q}}(v,w)_{[H]}$ containing the point $\varphi_t^{X_f}(q)$ and hence $\varphi_t^{X_f}(q) \in Rep_{\overline{Q}}(v,w)_H$. Therefore, $\varphi_t^{X_f}(\Phi^{-1}(\mathcal{O}_{\lambda_{\mathcal{K}^\star}})_{(H)}) \subseteq \Phi^{-1}(\mathcal{O}_{\lambda_{\mathcal{K}^\star}})_{(H)}$. So the Hamiltonian derivation $-ad_{f_{\lambda_{\mathcal{K}^\star}}}$ maps the symplectic piece or stratum $Rep_{\overline{Q}}^{\lambda_{\mathcal{K}^\star}}(v,w)_{(H)}$ of $Rep_{\overline{Q}}^{\lambda_{\mathcal{K}^\star}}(v,w)$ onto itself.

To prove (β), observe that $\Phi^{-1}(\mathcal{O}_{\lambda_{\mathcal{K}^\star}})_{(H)}$ is a K-invariant symplectic submanifold of $(Rep_{\overline{Q}}(v,w),\omega)$, and so the inclusion map

$$\iota_\Phi : \Phi^{-1}(\mathcal{O}_{\lambda_{\mathcal{K}^\star}})_{(H)} \longrightarrow Rep_{\overline{Q}}(v,w)$$

induces homomorphism of Poisson algebras

$$\begin{aligned} \iota_\Phi^* : (C^\infty(Rep_{\overline{Q}}(v,w),\omega)^K, \{\cdot,\cdot\}|_{\Phi^{-1}(\mathcal{O}_{\lambda_{\mathcal{K}^\star}})_{(H)}}, \cdot) \\ \downarrow \\ (C^\infty(\Phi^{-1}(\mathcal{O}_{\lambda_{\mathcal{K}^\star}})_{(H)},\omega)^K, \{\cdot,\cdot\}|_{\Phi^{-1}(\mathcal{O}_{\lambda_{\mathcal{K}^\star}})_{(H)}}, \cdot). \end{aligned} \tag{3.3}$$

Let $f_{\lambda_{\mathcal{K}^\star}}, g_{\lambda_{\mathcal{K}^\star}} \in C^\infty(Rep_{\overline{Q}}^{\lambda_{\mathcal{K}^\star}}(v,w),\mathbb{C})$, then, there exists

$f, g \in C^{\infty}(Rep_{\overline{Q}}(v,w), \mathbb{C})^K$ so that

$$\pi_R^*(f_{\lambda_{\mathcal{K}^*}}|Rep_{\overline{Q}}^{\lambda_{\mathcal{K}^*}}(v,w)_{(H)}) = f|\Phi^{-1}(\mathcal{O}_{\lambda_{\mathcal{K}^*}})_{(H)}$$

and

$$\pi_R^*(g_{\lambda_{\mathcal{K}^*}}|Rep_{\overline{Q}}^{\lambda_{\mathcal{K}^*}}(v,w)_{(H)}) = g|\Phi^{-1}(\mathcal{O}_{\lambda_{\mathcal{K}^*}})_{(H)}.$$

Thus

$$\begin{aligned}\{f_{\lambda_{\mathcal{K}^*}}|Rep_{\overline{Q}}^{\lambda_{\mathcal{K}^*}}(v,w)_{(H)},\, g_{\lambda_{\mathcal{K}^*}}|Rep_{\overline{Q}}^{\lambda_{\mathcal{K}^*}}(v,w)_{(H)}\}_{\lambda_{\mathcal{K}^*}}(\pi_{(q)}) =\\ = \{f|\Phi^{-1}(\mathcal{O}_{\lambda_{\mathcal{K}^*}})_{(H)},\, g|\Phi^{-1}(\mathcal{O}_{\lambda_{\mathcal{K}^*}})_{(H)}\}(q)\\ = \{f,\, g\}(q)\\ = \{f_{\lambda_{\mathcal{K}^*}},\, g_{\lambda_{\mathcal{K}^*}}\}_{\lambda_{\mathcal{K}^*}}(\pi_R(q)),\end{aligned}$$

for every $q = (\varphi, a, b) \in \Phi^{-1}(\mathcal{O}_{\lambda_{\mathcal{K}^*}})_{(H)}$. This shows that

$$\iota_{R^{\lambda_{\mathcal{K}^*}}} : Rep_{\overline{Q}}^{\lambda_{\mathcal{K}^*}}(v,w)_{(H)} \longrightarrow Rep_{\overline{Q}}^{\lambda_{\mathcal{K}^*}}(v,w)$$

is a Poisson map. The extension to the action of the complex reductive Lie group $G = K^{\mathbb{C}}$ follows the same arguments and techniques developed in R. Sjamaar [22] for the proof of the Holomorphic Slice Theorem. This completes the proof of the singular Reduction Theorem. ◊

4 An Important Example of the Construction of Quiver Varieties: Cotangent Bundles of Generalized Flag Manifolds of Type A_n

This example was intensively studied by several authors such as H. Nakajima [section 7, [18]] and also by G. Lusztig[section 1, [13]] in the case of the regular value of the momentum map induced by the proper and Hamiltonian action of the compact Lie group $K = \prod_{k \in Q_0} Aut(V_k)$ on the cotangent bundles of generalized flag manifolds of type A_n. We want investigate in this study the case of the singular value of the momentum map induced by the symplectic action of the reduced complex Lie group G on the generalized flag manifold of type A_n and then apply the singular reduction theorem (see theorem 3.1, section 3 above) in this case. The notations used here, will be the same as those in [22], [13] and section 2 of this article.
Let the quiver $Q := (Q_0, Q_1, s, t : Q_1 \longrightarrow Q_0)$ be the Dynkin diagram of type $A_n, n \geq 1$, i.e.

$$\overset{1}{\bullet} \xrightarrow{h_1} \overset{2}{\bullet} \xrightarrow{h_2} \overset{3}{\bullet} \xrightarrow{h_3} \cdots \xrightarrow{h_n} \overset{n}{\bullet} \qquad (4.1)$$

where as in section 2 above $s(h_i)$ and $t(h_i)$ are respectively the source and the target of the arrows h_i in the set of arrows Q_1 for any vertices $i = 1, 2, \cdots, n$ in the set of vertices Q_0. For more details see the references cited in this section. The generalized Cartan matrix associated to the quiver Q is defined by $C = 2I - A_Q$ with the adjacency matrix $A_Q = (a_{k,l})_{k,l \in Q_0}$, and with

$$a_{kl} = \begin{cases} 0, & if\, k \neq l \\ 1, & if\, k = l \end{cases}$$

the number of arrows between vertices k and l in this order. Let $V = (V_k)_{k\in Q_0}$ be collection of finite dimensional $\mathbb{C}$-vector spaces attached to each vertex $k \in Q_0$, and r a positive integer such that there exists a strictly decreasing sequence of integers $(v_k)_{k\in Q_0}$ defined as follows:

$$r > v_1 > v_2 > \cdots > v_n > 0.$$

Following [18], [13], and section 2 of this article, let $\dim V = v = (v_1,\cdots,v_n)$, and $\dim W = w = (r,0,\cdots,0)$ be elements of $\mathbb{Z}^{Q_0}_{\geq 0}$ and $(\varphi)_{h\in Q_1}$ elements of the variety $\bigoplus_{h\in Q_1} Hom(V_{s(h)},V_{t(h)})$, $a \in Hom(W,V)$ and $b \in Hom(V,W)$. Define a generalized *flag* manifold of type A_n by

$$\mathcal{F}(v,r) := \{\mathcal{F} = \{\mathbb{C}^r = V^0 \supset V^1 \supset \cdots \supset V^n \supset V^{n+1} = \{0\}\}\}, \tag{4.2}$$

to be a finite dimensional smooth complex manifold consisting of all flags $\mathcal{F}$ that are sequences of Q_0- graded subspaces of $\mathbb{C}^r$, i.e., $\varphi_h(V^{l-1}_{s(h)}) \subset V^l_{t(h)}$, such that $\dim V^l = v_l$, for any $l \in Q_0$ and assume $W = V^0$, for more details see [13, Lemma 1.8, (a)]. Hence, the stability of the variety $\mathcal{F}(v,r)$ is deduced from the stability of the vector spaces of V. Fix the orientation $\Omega \subset Q_1$, and define the variety

$$\mathcal{F}_\Omega(\varphi,\mathcal{F}) := \{(\varphi,\mathcal{F}) \in \bigoplus_{h\in\Omega} Hom(V_{s(h)},V_{t(h)}) \times \mathcal{F}(v,r) : \\ \varphi_h(V^{k-1}) \subset V^k, \ \forall h \in Q_1\},$$

This is a finite dimensional $\mathbb{C}$-vector space, with dimension

$$\dim_{\mathbb{C}} \mathcal{F}_\Omega(\varphi,\mathcal{F}) = \sum_{l'\leq l} \sharp\{h \in \Omega : s(h) = l', t(h) = l\} v_{l'} v_l + \sum_{l<l'} v_{l'} v_l,$$

for more details see [13, Lemma 1.6, (c)]. Then the variety

$$\mathcal{F}_{Q_1}(\varphi,\mathcal{F}) := T^\star \mathcal{F}_\Omega(\varphi,\mathcal{F})$$

is the cotangent bundle of the the variety $\mathcal{F}_\Omega(\varphi,\mathcal{F})$ and its finite dimensional $\mathbb{C}$-vector space is:

$$\dim_{\mathbb{C}} \mathcal{F}_{Q_1}(\varphi,\mathcal{F}) = \sharp_{l'\leq l} \{h \in Q_1 : s(h) = l', t(h) = l\} v_{l'} v_l + 2\sum_{l<l'} v_{l'} v_l.$$

Hence, the variety $\mathcal{F}_{Q_1}(\varphi,\mathcal{F})$ is a symplectic manifold with the closed 2-form ω. The proper and Hamiltonian action of the compact Lie group $K = \prod_{k\in Q_0} Aut(V_k)$ on the generalized flag manifold $\mathcal{F}(v,r)$ given by $(k,\mathcal{F}) \mapsto k\mathcal{F} = (k_{t(h)}\mathcal{F}k^{-1}_{s(h)})_{h\in Q_1} = ((k_{t(h)}Vk^{-1}_{s(h)}) = (k_{t(h)}V^0k^{-1}_{s(h)}) \supset (k_{t(h)}V^1k^{-1}_{s(h)}) \supset \cdots \supset (k_{t(h)}V^{n+1}k^{-1}_{s(h)}) = 0, \ \forall h \in Q_1)$. Thus G acts on the cotangent bundles $\mathcal{F}_{Q_1}(\varphi,\mathcal{F})$ by $(g,(\varphi,\mathcal{F})) \mapsto (g\varphi,g\mathcal{F})$ giving rise to the momentum map $\Phi_{\mathbb{C}}$: $\mathcal{F}_{Q_1}(\varphi,\mathcal{F}) \longrightarrow \mathcal{G}^\star$ which is in turn given by the complex ADHM equations defined by

$(\varphi,\mathcal{F}) \mapsto \Phi_{\mathbb{C}}(\varphi,\mathcal{F}) = \varepsilon\varphi\varphi + ab = \lambda = (\lambda_{\mathcal{K}^\star},\lambda_{\mathbb{C}})$, where $\lambda_{\mathcal{K}^\star}$ is the real component of λ which is the value of the momentum map induced by the action of K on $\mathcal{F}_{Q_1}(\varphi,\mathcal{F})$ with $\mathcal{K}^\star$ the dual of the Lie algebra $\mathcal{K} = \mathrm{Lie}K$, and $\lambda_{\mathbb{C}}$ the complex component of λ, where $\varepsilon : h \in Q_1 \mapsto \varepsilon(h) \in \{-1,1\}$. Following H. Nakajima [18], $\lambda_{\mathbb{C}} = 0$ implies that when the components of $\lambda_{\mathcal{K}^\star}$, i.e., $\lambda^k_{\mathcal{K}^\star} \geq 0$ for any $k \in Q_1$, then $\lambda = (\lambda_{\mathcal{K}^\star},0)$ is the regular value of the momentum map $\Phi_{\mathbb{C}}$. Therefore, the quotient $\Phi_{\mathbb{C}}^{-1}(\lambda)//_G$ is a smooth manifold, and its stable component $\Phi_{\mathbb{C}}^{-1}(\lambda)^s//_G$ is isomorphic to the cotangent bundle $\mathcal{F}_{Q_1}(\varphi,\mathcal{F})$. Hence the map $\Phi_{\mathbb{C}}^{-1}(0)//_G \ni (\varphi,a,b) \mapsto b_1a_1 \in End(\mathbb{C}^r)$ is proper.

Next, assume λ is singular and an orbitally convex closed coadjoint orbit $O_\lambda := O_{\lambda_{\mathcal{K}^\star}} \otimes_{\mathbb{R}} \mathbb{C}$ passing through λ, where $O_{\lambda_{\mathcal{K}^\star}}$ is the closed real coadjoint orbit passing through $\lambda_{\mathcal{K}^\star}$, then the quotient $\mathcal{F}^\lambda_{Q_1}(\varphi,\mathcal{F}) := \Phi_{\mathbb{C}}^{-1}(O_\lambda)//_G$ is not a manifold. H. Nakajima [18], and G. Lusztig [13], and also applying R. Sjamaar's Holomorhic Slice Theorem [22, Theorem 1.12] the complex quiver variety $\mathcal{F}^\lambda_{Q_1}(\varphi,\mathcal{F}) := \Phi_{\mathbb{C}}^{-1}(O_\lambda)//_G$ is a stratified symplectic space, and the stratification is the same as in section 2 of this article. So, for any point $[(\varphi,a,b)] \in \Phi_{\mathbb{C}}^{-1}(O_\lambda)//_G$, let $H^{\mathbb{C}} = \{g \in G : g(\varphi,a,b) = (\varphi,a,b)\} = H \otimes_{\mathbb{R}} \mathbb{C}$ be the stabilizer of (φ,a,b) in G which is the complexified of the stabilizer H of (φ,a,b) in K. Then the strata $\mathcal{F}^\lambda_{Q_1}(\varphi,\mathcal{F})_{(H^{\mathbb{C}})}$ are smooth symplectic manifolds for any stabilizer $H^{\mathbb{C}} \subset G$ of (φ,a,b) in $\mathcal{F}_{Q_1}(\varphi,\mathcal{F})$. So, for any point $[(\varphi,a,b)] \in \mathcal{F}^\lambda_{Q_1}(\varphi,\mathcal{F})$ there is a stabilizer $H^{\mathbb{C}} \subset G$ such that $[(\varphi,a,b)] \in \mathcal{F}^\lambda_{Q_1}(\varphi,\mathcal{F})_{(H^{\mathbb{C}})}$. Let the sequences of subspaces$\mathbb{C}^r = V^0 \supset V^1 \supset \cdots \supset V^n \supset V^{n+1} = 0$ be such that $V^k := Imb_1\varphi_{h_1}\varphi_{h_2}\cdots\varphi_{h_k}, \forall h_i \in Q_1, k \in Q_0$ and $1 \leq i \leq k$, this a finite dimensional $\mathbb{C}$-vector subspace of $\mathbb{C}^r$. The Poisson structure on $\mathcal{F}_{Q_1}(\varphi,\mathcal{F})$ induces the Poisson structure on the quiver variety $\mathcal{F}^\lambda_{Q_1}(\varphi,\mathcal{F})$, i.e., $\forall f_\lambda \in C^\infty(\mathcal{F}^\lambda_{Q_1}(\varphi,\mathcal{F}))$ there exists $f \in C^\infty(\mathcal{F}_{Q_1}(\varphi,\mathcal{F}))$ such that

$$\pi^\star_{\mathcal{F}} f_\lambda = f|_{\Phi_{\mathbb{C}}^{-1}(O_\lambda)},$$

where $\pi_{\mathcal{F}} : \mathcal{F}_{Q_1}(\varphi,\mathcal{F}) \longrightarrow \mathcal{F}^\lambda_{Q_1}(\varphi,\mathcal{F})$ is a canonical projection.

Let $x = (\varphi,a,b) \in \mathcal{F}_{Q_1}(\varphi,\mathcal{F})$ and $K_x \subset K$ its isotropy group. The stratifiaction of $\mathcal{F}_{Q_1}(\varphi,\mathcal{F}) = \sqcap_{K_x\subset K} \mathcal{F}_{Q_1}(\varphi,\mathcal{F})_{(K_X)}$ induces the stratification of $\mathcal{F}^{\lambda_{\mathcal{K}^\star}}_{Q_1}(\varphi,\mathcal{F}) = \sqcap_{K_x\subset K} \mathcal{F}^{\lambda_{\mathcal{K}^\star}}_{Q_1}(\varphi,\mathcal{F})_{(K_X)}$ with $\mathcal{F}^{\lambda_{\mathcal{K}^\star}}_{Q_1}(\varphi,\mathcal{F})_{(K_X)} = \pi_{\mathcal{F}}(\mathcal{F}_{Q_1}(\varphi,\mathcal{F})_{(K_X)} \cap \Phi^{-1}(O_{\lambda_{\mathcal{K}^\star}}))$ with $\Phi : \mathcal{F}_{Q_1}(\varphi,\mathcal{F}) \longrightarrow \mathcal{K}^\star$ the momentum map of the action of K on $\mathcal{F}_{Q_1}(\varphi,\mathcal{F})$. So, the stratum $\mathcal{F}^{\lambda_{\mathcal{K}^\star}}_{Q_1}(\varphi,\mathcal{F})_{(K_X)}$ has smooth structures. Therefore, each $\mathcal{F}^{\lambda_{\mathcal{K}^\star}}_{Q_1}(\varphi,\mathcal{F})_{(K_X)}$ satisfies the conditions of the theorem 3.1. Next, let K_x and K_y be respectively isotropy subgroups of K at x and y points of $\mathcal{F}^{\lambda_{\mathcal{K}^\star}}_{Q_1}(\varphi,\mathcal{F})_{(K_X)}$ such that $K_x \subset K_y$ and the class of K_x are in the class of K_y, i.e. $(K_x) \subset (K_y)$, then, following M.J.Plaum [21] and R. Sjamaar [22] the frontier property of the stratification conditions states $\mathcal{F}^{\lambda_{\mathcal{K}^\star}}_{Q_1}(\varphi,\mathcal{F})_{(K_y)} \cap \overline{\mathcal{F}^{\lambda_{\mathcal{K}^\star}}_{Q_1}(\varphi,\mathcal{F})}_{(K_x)}$ is a nonempty manifold and it satisfies the conditions of the Theorem 3.1. Therefore, by the extension of the symplectic action of the complex reductive Lie group G on $\mathcal{F}_{Q_1}(\varphi,\mathcal{F})$, then the reduced space $\mathcal{F}^\lambda_{Q_1}(\varphi,\mathcal{F})$ satisfies the conditions of the Theorem 3.1.

References

[1] L. Bates and E. Lerman, Proper Group Actions and Symplectic Stratified Spaces, *Pacific J. Math.* **181** (1997), n0. 0, 201-229.

[2] R. Cushman, Lectures on Reduction Theory for Hamiltonian Systems, *The Spring School at the Mathematics Institute of Utrecht*, **3-25** June 2004.

[3] K. Fritzche and H.Grauert, From Holomorphic Functions to Complex Manifolds. *Graduate Texts in Mathematics,* **213**. Springer-Verlag, Berlin, New-York, 2002.

[4] V. Ginzburg, Non-Commutative Geometry, Quiver Varieties and Operads, *Math. Res. Lett.* **8** (2001), no. 3, 377-400.

[5] M. Goresky and R. MacPherson, *Stratified Morse Theory*, Springer-Verlag, New York (1988).

[6] P. Heinzner, Geometric Invariant Theory on Stein Spaces, *Math. Ann.* **289** (1991), no. 4, 631-662.

[7] N. J. Hitchin, A. Karlhde, U. Lindstrom, and M. Rocek, Hyper-Kähler Metrics and Super Symmetry, *Comm. Math. Physics* **108** (1987), no. 4, 535-589.

[8] F. Kirwan, Cohomology of Quotients in Symplectic and Algebraic Geometry, *Mathematicla Notes* **31**. Princeton University Press, Princeton, NJ, 1984.

[9] A. King, Moduli of Representations of Finite Dimensional Algebras, *Quart. J. Math. Oxford Ser.* **(2)** 45 (1994), no. 180, 515-530.

[10] E. Lerman, R. Montgomery and R. Sjamaar, Examples of Singular Reduction, 127-155, *London Math. Soc. Lecture Note Ser.*, **192**, Cambridge Univ. Press, Cambridge, 1993.

[11] E. Lerman and R. Sjamaar, Reductive Group Action on Kähler Manifolds, Conservative Systems and Quantum Chaos (Waterloo, Ontario, 1992), 85-92, *Fields Inst. Commun.*, **8**, *Amer. Math. Soc.,* Providence, RI, 1996.

[12] G. Lusztig, On Quiver Varieties, *Adv. Math.* **136** (1998), 141-182.

[13] G. Lusztig, Quivers, Perverse Sheaves and Quantized Enveloping Algebras, *J. Amer. Math. Soc.* **4**(1991), no. 2, 365-421.

[14] D. Luna, Slices Etales. (French) Etales sur les groupes algébriques, pp. 81-105. *Bull. Soc. Math.* France, Paris, *Memoire* **33** *Soc. Math.* France, Paris, 1973.

[15] C. M. Marle, Modele d'Action Hamiltonianne d'un Groupe de Lie sur une Variete Symplectique, *Rend. Sem. Mat. Univ. Politec. Torino* **43**(1985), no.2, 227-251 (1986).

[16] J. Marsden and A. Weinstein , Reduction of Symplectic Manifolds with Symmetry, *Rep. Mathematical. Phys.* **5** (1974), no. 1, pp. 121-130.

[17] H. Nakajima, Quiver Varieties and Kac-Moody Algebras, *Duke Math. J.* **91**(3)(1998), no. 3, 515-560.

[18] H. Nakajima, Instantons on ALE Spaces, Quiver Varieties and Kac-Moody Algebras, *Duke Math. J.* **76**(1994), no. 2, 365-416.

[19] H. Nakajima, Quiver Varieties and Finite Dimensional Representations of Quantum Affine Algebras, *J. Amer. Math. Soc.* **14**(2000), no. 1, 145-238.

[20] R. S. Palais, On the Existence of Slices for Actions of Non-Compact Lie Groups, *Ann. of Math,* **73** (1961), 295-323.

[21] M. J. Pflaum, Analytic and Geometric Study of Stratified Spaces, *Lecture Notes in Mathematics*, **1768**. Springer-Verlag, Berlin, (2001).

[22] R. Sjamaar, Holomorphic Slices, Symmetric Reduction and Multiplicities of Representations, *Ann. of Math.* **(2)** 141 (1995), no. 1, 87-129.

[23] R. Sjamaar and E. Lerman, Stratified Symplectic Spaces and Reductions, *Ann.of Math.* **(2)** 134 (1991), no. 2, 375-422.

[24] J. Śniatycki, *Orbits of families of Vector fields on subcartesian spaces* , (to appear in Annales Inst. Fourier (Grenoble)), arxiv.math.DG./0211212.

[25] D. M. Snow, Reduction Group Actions on Stein Spaces, *Math. Ann.* **259** (1982), no. 1, 79-97.

In: African Diaspora Mathematics Research Progress
Editor: Toka Diagana, pp. 129–145
ISBN 978-1-60456-204-0
© 2008 Nova Science Publishers, Inc.

Chapter 9

MATHIEU FUNCTION AND KONTOROVICH-LEBEDEV TRANSFORMS IN THE L-SHAPED WAVE SCATTERING PROBLEM

L. P. Castro *
Department of Mathematics, University of Aveiro
3810–193 Aveiro, Portugal
A. H. Kamel †
PO Box 433 Heliopolis Center 11757, Cairo, Egypt

Abstract

We consider a boundary-value problem for the Helmholtz equation outside a right-angled wedge configuration formed by a half-plane and a strip (i.e., the so-called L-shaped surface boundary). The problem models the diffraction of plane waves by scatterers of such L-shaped configurations. The proposed scheme for the solution of the problem includes an application of the Kontorovich-Lebedev (KL) transform and a new discrete index of the Mathieu function (diMf) transform. Within the present approach, an integral equation satisfied by the KL spectrum, and a linear system for the diMf spectral amplitudes are derived. In addition, the singularities of the spectral function are deduced. Moreover, near and far field representations are also obtained.

AMS Subject Classification: 30E25, 33E10, 35J05, 44A15, 45H05.

Keywords: Wave scattering, boundary-value problem, Helmholtz equation, Kontorovich-Lebedev transform, Mathieu function.

1 Introduction

Scalar wave propagation in a space imbedded with an L-shaped surface boundary is an important topic in diffraction theory, and relevant for many engineering applications. From the mathematical point of view, such situation can be stated as a particular boundary-value

*E-mail: lcastro@mat.ua.pt
†E-mail: alaa@ureach.com

problem for the Helmholtz equation. In other geometrically simpler situations it is well-known how to apply known integral transforms or series expansions, and derive corresponding rigorous solutions. For some other situations, although the closed form solutions are not known, eigenfunctions and asymptotic solutions have been obtained.

In the present L-shaped geometry, a technique based on integral operators of Wiener–Hopf–Hankel type was recently proposed in [1] for the Dirichlet–Neumann boundary conditions case, and using only a non-real wave number (due to a lossy medium). There, the final representation of the solution was provided upon certain factorization schemes involving semi-almost periodic matrix-valued functions (cf. also [2] for a part of the known theory about such kind of factorizations). The techniques used in [1] depend strongly on the fact of using a non-real wave number. The methods proposed in the present paper bring out an alternative formulation of such problems in terms of KL and diMf transforms, leading to additional detailed conclusions which hold true in both cases of pure real and complex (non-real) wave numbers.

We should clarify that Mathieu equations occur naturally in applications involving elliptic geometries, as well as in problems involving periodic motion (like the trajectory of an electron in a periodic array of atoms or the mechanics of the quantum pendulum). In here, the initial motivation arose from the first case by using elliptic coordinates and obtaining the well-known separation of the wave equation.

The structure of the present paper is as follows. The problem is formulated in Section 2. In Section 3, the integral equation satisfied by the KL spectrum and the linear system associated with the diMf spectra are derived. Field representations are obtained in Section 4. Conclusions are assembled in Section 5. Additionally, an Appendix A is added to derive the diMf transform, and in the Appendix B is summarized the KL integral transform.

2 Formulation

We consider the problem of scattering of harmonic acoustic waves by an L-shaped surface boundary imbedded in an infinite medium. We should mention from the very beginning that the analysis of the acoustic waves carried out here is not restrictive. It applies to any other type of scalar waves including transverse electric and transverse magnetic waves of electromagnetism, as well as shear horizontal waves of elastodynamics.

The formulation here presented handles the following four boundary situations: Dirichlet(D)–Neumann(N), N–D, N–N and D–D boundary conditions on the L-shaped surface formed by an half-plane and a strip. We will deal with the prototype D–D case, without loss of generality.

In what follows (u,v,z) denote the usual elliptic cylindrical coordinates such that the strip connects the right-hand focus of the elliptic coordinate system with the origin. The region $C_1 = \{(u,v,z)\,|\,0<u<\infty, 0\leq v<\pi/2, -\infty<z<\infty\} \cup \{(u,v,z)\,|\,0<u<\infty, 3\pi/2< v<2\pi, -\infty<z<\infty\}$ defines the right-hand side of the space.

Let (r,ϕ,z) denote the usual cylindrical coordinates such that on the half-plane one has $\phi=\pi/2$. We will denote by $C_2 = \{(r,\phi,z)\,|\,0<r<\infty, \pi/2<\phi<3\pi/2, -\infty<z<\infty\}$ the left-half side of the space.

A time factor $\exp(-i\omega t)$ is assumed and omitted throughout the paper. The parameters k, ρ and λ will denote the wave number, density and incompressibility of the space, respectively.

The acoustic field is describable by the pressure p and velocity $\mathbf{V}$, obeying to the Euler field equations [3]

$$\frac{-i\omega}{\lambda}p+\nabla.\mathbf{V}=-s\,, \tag{2.1}$$

$$\nabla p-i\omega\rho\mathbf{V}=-\mathbf{f}\,. \tag{2.2}$$

The excitation terms s and $\mathbf{f}$ represent the scalar particle source and the impressed vector force densities, respectively; ∇ is the spatial gradient operator.

In what follows s is assumed to be zero; acoustic field excitation is provided by an impressed unit line force density located at (r_0,ϕ_0) in C_2. The linearity of the field equations in (2.1) and (2.2) implies that fields excited by $z-$independent but otherwise arbitrary impressed force density can be expressed in terms of linear superposition. The situation when the sources are located in C_1 can be described in a similar way to the one to be detailed here.

2.1 On the C_1 Region

With ρ and λ being assumed constants, and from Eqs. (2.1)–(2.2), the acoustic pressure in C_1, $p_1(u,v)$, satisfies the homogeneous Helmholtz equation

$$(\nabla^2+k^2)p_1(u,v)=0\,, \tag{2.3}$$

where

$$\nabla^2=\frac{1}{h^2}\left[\frac{\partial^2}{\partial u^2}+\frac{\partial^2}{\partial v^2}\right], \tag{2.4}$$

and $k=\omega/c$, $c=\sqrt{\lambda/\rho}$ is the acoustic speed, $h^2=w^2(\cosh^2 u-\cos^2 v)$ is the Jacobian of the transformation from the (x,y)–coordinate system to the (u,v)–coordinate system and $2w$ is the distance between the two foci of the elliptic system.

The velocity fields are derived from the acoustic pressures as

$$\mathbf{V}_1(u,v)=\frac{-i}{\omega\rho h}\left[\mathbf{u}_0\frac{\partial}{\partial u}+\mathbf{v}_0\frac{\partial}{\partial v}\right]p_1(u,v)\,. \tag{2.5}$$

The fields are required to satisfy the Sommerfeld radiation condition, as $u\longrightarrow\infty$.

Following the condition of Meixner [4], we assume that the energy stored in any finite neighborhood of the corner of the L-shaped surface boundary must be finite, that is,

$$\int_S\left(\rho|\mathbf{V}|^2+\frac{1}{\rho c^2}p^2\right)dS\longrightarrow 0 \tag{2.6}$$

as the surface S contracts to the neighborhood of the corner.

Soft boundary Dirichlet conditions are imposed on the surface of the strip and on the half-plane, namely

$$p_1(u=0,v\in C_1)=0\,, \tag{2.7}$$

$$p_1\left(\infty>u\geq 0,v=\tfrac{\pi}{2}\right)=0\,, \tag{2.8}$$

respectively.

We propose to represent the field in C_1 in terms of a diMf transform (see Appendix A).

Theorem 2.1 *In the C_1 region, we have the representation formula*

$$p_1(u,v) = \sum_p B(\nu_p) f_{\nu_p}(kw,v)\Phi_p(kw,u) , \tag{2.9}$$

where $B(\nu_p)$ is the diMf transform spectra,

$$f_{\nu_p}(kw,v) = C_{\nu_p}(kw,\pi/2)S_{\nu_p}(kw,v) - S_{\nu_p}(kw,\pi/2)C_{\nu_p}(kw,v) ,$$

$$S_{\nu_p}(kw,v) = \sum a_n \sin[(\upsilon_p+2n)v] ,$$

$$C_{\nu_p}(kw,v) = \sum b_n \cos[(\upsilon_p+2n)v] ,$$

and Φ_p being given in (A.9).

Applying the index transform

$$p_1(u,v) = \sum_p P_1(\nu_p,v)\Phi_p(kw,u) \tag{2.10}$$

to Eq. (2.3), and making use of the orthogonality relation in Eq. (A.10), we obtain the ordinary differential equation (ODE)

$$\left[\frac{d^2}{d\phi^2} + (\nu_p^2 - k^2w^2\cos^2 v)\right] P_1(\nu_p,v) = 0 \tag{2.11}$$

which solution is given in terms of angular Mathieu functions [5].

In what follows we choose for the ODE in Eq. (2.11) solutions of the form

$$S_{\nu_p}(kw,v) = \sum a_n \sin[(\upsilon_p+2n)v] , \tag{2.12}$$

$$C_{\nu_p}(kw,v) = \sum b_n \cos[(\upsilon_p+2n)v], \tag{2.13}$$

$$f_{\nu_p}(kw,v) = C_{\nu_p}(kw,\pi/2)S_{\nu_p}(kw,v) - S_{\nu_p}(kw,\pi/2)C_{\nu_p}(kw,v) . \tag{2.14}$$

Thus

$$P_1(\nu_{1p},v) = B(\nu_p) f_{\nu_p}(kw,v) , \tag{2.15}$$

where $B(\nu_p)$ is the diMf transform spectra to be determined from the *joining boundary conditions* – cf. the corresponding subsection bellow. Hence, from (2.10) and (2.15), we obtain

$$p_1(u,v) = \sum_p B(\nu_p) f_{\nu_p}(kw,v)\Phi_p(kw,u) . \tag{2.16}$$

As about the velocity field, it is now derived from Eq. (2.5) by using the p_1 formula presented in the last theorem. Finally, relevant to our problem is also the angular velocity field component

$$\mathbf{V}_{1v}(u,v) = \frac{-i}{\omega\rho h} \sum_p B(\nu_p) \frac{d}{dv} f_{\nu_p}(kw,v)\Phi_p(kw,u) . \tag{2.17}$$

2.2 On the C_2 Region

The acoustic pressure in C_2, $p_2(r,\phi)$, satisfies the inhomogeneous Helmholtz equation

$$(\nabla^2+k^2)p_2(r,\phi) = -\frac{\delta(r-r_0)\,\delta(\phi-\phi_0)}{r_0}\,. \tag{2.18}$$

We propose to represent the acoustic pressure in C_2 in terms of a Kontorovich-Lebedev (KL) transform [6]. For such a purpose, let us first recall the definition of the Kontorovich-Lebedev transform:

$$P(\nu,\phi) = \int_0^\infty \frac{1}{r} p(r,\phi) H_\nu^{(1)}(kr)dr\,, \tag{2.19}$$

$$p(r,\phi) = \frac{1}{2}\int_{-i\infty}^{i\infty} \nu J_\nu(kr) P(\nu,\phi)d\nu\,, \tag{2.20}$$

where ν is a pure imaginary number, and $J_\nu(z)$ and $H_\nu^{(1)}(z)$ are the standard Bessel and Hankel functions, respectively.

Theorem 2.2 *In the C_2 region, we have*

$$p_2(r,\phi) = \frac{-i}{2}\int_0^{i\infty} \nu \exp(i\pi\nu)\,\sin\pi\nu\, H_\nu^{(1)}(kr)\, P_2(\nu,\phi)\, d\nu\,. \tag{2.21}$$

Since (cf. [3])

$$H_{-\nu}^{(1)}(z) = \exp(i\pi\nu)\, H_\nu^{(1)}(z)\,, \tag{2.22}$$

the definition in Eq. (2.19) implies that

$$P(-\nu,\phi) = \exp(i\pi\nu)\, P(\nu,\phi)\,. \tag{2.23}$$

Moreover, there are two additional ways of writing the inverse transform of Eq. (2.20):
a) From Eqs. (2.22)–(2.23) and making use of

$$J_\nu(z) = \tfrac{1}{2}[H_\nu^{(1)}(z) + H_\nu^{(2)}(z)]\,, \tag{2.24}$$

$$H_{-\nu}^{(2)}(z) = \exp(-i\pi\nu)\, H_\nu^{(2)}(z)\,, \tag{2.25}$$

we obtain

$$p(r,\phi) = \frac{1}{4}\int_{-i\infty}^{i\infty} \nu H_\nu^{(1)}(kr) P(\nu,\phi)d\nu\,. \tag{2.26}$$

b) Using now Eqs. (2.22)–(2.23), we have

$$p(r,\phi) = \frac{-i}{2}\int_0^{i\infty} \nu \exp(i\pi\nu)\,\sin\pi\nu\, H_\nu^{(1)}(kr) P(\nu,\phi)d\nu\,, \tag{2.27}$$

which allows the statement.

Similarly as in the former case, the fields are required to satisfy the Sommerfeld radiation condition as $r \to \infty$. Following the condition of Meixner, the field behavior near the corner of the L-shaped surface boundary is

$$p = O(r^\nu), \quad |\mathbf{V}| = O(r^{\nu-1}), \quad r \to 0, \quad \nu > 0\,. \tag{2.28}$$

With the knowledge of Theorem 2.2, the velocity fields in C_2 are now derived from the acoustic pressures as

$$\mathbf{V}_2(r,\phi) = \frac{-i}{\omega\rho}\left[\mathbf{r}_0\frac{\partial}{\partial r} + \phi_0\frac{1}{r}\frac{\partial}{\partial\phi}\right] p_2(r,\phi)\,. \tag{2.29}$$

Relevant to the problem under consideration is also the velocity component $V_{2\phi}(r,\phi)$ normal to the half-plane, namely

$$V_{2\phi}(r,\phi) = \frac{-1}{2\omega\rho}\frac{1}{r}\frac{\partial}{\partial\phi}\int_0^{i\infty} \nu\exp(i\pi\nu)\sin\pi\nu\, H_\nu^{(1)}(kr)P_2(\nu,\phi)\,d\nu\,. \tag{2.30}$$

Therefore, one obtains

$$V_{2\phi}(r,\phi) = \frac{-1}{\omega\rho}\frac{1}{r}\int_0^{i\infty} \nu\exp(i\pi\nu)\,\sin\pi\nu\, H_\nu^{(1)}(kr)\frac{d}{d\phi}P_2(\nu,\phi)\,d\nu\,. \tag{2.31}$$

It is interesting to observe that, by applying now the KL-transform to Eq. (2.18), we get the ordinary differential equation

$$\left[\frac{d^2}{d\phi^2}+\nu^2\right] P_2(\nu,\phi) = -H_\nu^{(1)}(kr_0)\delta(\phi-\phi_0)\,, \quad \frac{\pi}{2}<\phi<\frac{3\pi}{2}\,. \tag{2.32}$$

We will represent from now on the field in C_2 as

$$p_2(r,\phi) = p_2^{(0)}(r,\phi) + p_2^{(1)}(r,\phi), \tag{2.33}$$

leading to

$$P_2(\nu,\phi) = P_2^{(0)}(\nu,\phi) + P_2^{(1)}(\nu,\phi)\,, \tag{2.34}$$

and in the next two subsections we will analise these last two terms.

2.2.1 The $P_2^{(0)}(\nu,\phi)$ Field

We choose the $P_2^{(0)}(\nu,\phi)$ field to satisfy the source conditions of Eq. (2.32), namely

a) $P_2^{(0)}(\nu,\phi)$ is continuous across $\phi=\phi_0$

$$P_2^{(0)}(\nu,\phi_0-0) = P_2^{(0)}(\nu,\phi_0+0)\,, \tag{2.35}$$

b) $\frac{d}{d\phi}P_2^{(0)}(\nu,\phi)$ is discontinuous across $\phi=\phi_0$

$$\frac{d}{d\phi}P_2^{(0)}(\nu,\phi)_{|\phi=\phi_0+0} - \frac{d}{d\phi}P_2^{(0)}(\nu,\phi)_{|\phi=\phi_0-0} = -H_\nu^{(1)}(kr_0)\,. \tag{2.36}$$

Proposition 2.3 *Within the last conditions, it holds*

$$P_2^{(0)}(\nu,\phi) = \frac{H_\nu^{(1)}(kr_0)\cos(\nu[\frac{3\pi}{2}-\phi])\sin(\nu[\phi_0-\frac{\pi}{2}])}{\nu\cos(\nu\pi)}\,, \qquad \phi>\phi_0, \tag{2.37}$$

$$P_2^{(0)}(\nu,\phi) = \frac{H_\nu^{(1)}(kr_0)\cos(\nu[\frac{3\pi}{2}-\phi_0])\sin(\nu[\phi-\frac{\pi}{2}])}{\nu\cos(\nu\pi)}, \qquad \phi<\phi_0. \tag{2.38}$$

Additionally, $V_{2\phi}^{(0)}(r,\phi)$ *is therefore given by*

$$V_{2\phi}^{(0)}(r,\phi) = \frac{-1}{\omega\rho}\frac{1}{r}\int_0^{i\infty}\nu\exp(i\pi\nu)\sin\pi\nu H_\nu^{(1)}(kr)\frac{d}{d\phi}P^{(0)}(\nu,\phi)\,d\nu, \tag{2.39}$$

where we have for $\phi>\phi_0$

$$\frac{d}{d\phi}P_2^{(0)}(\nu,\phi) = \frac{H_\nu^{(1)}(kr_0)\sin(\nu[\frac{3\pi}{2}-\phi])\sin(\nu[\phi_0-\frac{\pi}{2}])}{\cos(\nu\pi)}, \tag{2.40}$$

and for $\phi<\phi_0$

$$\frac{d}{d\phi}P_2^{(0)}(\nu,\phi) = \frac{H_\nu^{(1)}(kr_0)\cos(\nu[\frac{3\pi}{2}-\phi_0])\cos(\nu[\phi-\frac{\pi}{2}])}{\cos(\nu\pi)}. \tag{2.41}$$

The formulas are obtained by a direct use of the two conditions in Eqs. (2.35) and (2.36), and from the soft boundary requirement at $\phi=\pi/2$.

2.2.2 The $P_2^{(1)}(\nu,\phi)$ Field

We turn now to the analysis of the $P_2^{(1)}$ field.

Proposition 2.4 *We have the identity*

$$P_2^{(1)}(\nu,\phi) = A(\nu)\sin\left(\nu\left[\phi-\frac{\pi}{2}\right]\right), \tag{2.42}$$

where

$$A(-\nu) = -\exp(i\pi\nu)A(\nu), \tag{2.43}$$

and

$$A(\nu) = O\left[\exp\left(-\frac{3\pi}{2}\Im m\nu\right)\right], \tag{2.44}$$

when $\Im m\nu\to+\infty$.

Since $P_2(\nu,\phi)$ must satisfy the soft boundary (Dirichlet) condition at $\phi=\pi/2$, we can represent it by (2.42), where $A(\nu)$ is a KL spectrum to be determined from the *joining conditions* (cf. the next subsection).

Additionally, from Eq. (2.42), Eq. (2.23) enforces the identity (2.43).

Finally, the convergence of the KL integrals [7] at $\phi=3\pi/2$ implies that, when $\Im m\nu\to+\infty$, the spectral functions must vanish as (2.44).

From the above we also obtain

$$p_2^{(1)}(r,\phi) = \frac{-i}{2}\int_0^{i\infty}\nu\exp(i\pi\nu)\sin\pi\nu H_\nu^{(1)}(kr)A(\nu)\sin(\nu[\phi-\frac{\pi}{2}])\,d\nu,$$

$$V_{2\phi}^{(1)}(r,\phi) = \frac{-1}{2\omega\rho}\frac{1}{r}\int_0^{i\infty}\nu^2\exp(i\pi\nu)\sin\pi\nu H_\nu^{(1)}(kr)A(\nu)\cos(\nu[\phi-\frac{\pi}{2}])\,d\nu.$$

2.3 The Joining Boundary Conditions

The field representation in the two regions C_1 and C_2 are jointed by imposing the both requirements of the continuity of the acoustic pressure and the normal velocity on the half-plane $v = \phi = 3\pi/2$. Putting now all conditions together, we directly obtain the identities assembled in the next result.

Corollary 2.5

$$\frac{-i}{2}\int_0^{i\infty} \nu \exp(i\pi\nu)\sin\pi\nu H_\nu^{(1)}(kr)[F_1(\nu)+A_1(\nu)\sin(\nu\pi)]\,d\nu \\ = \sum_p B(\nu_p) f_{\nu_p}(kw,v)_{|v=3\pi/2}\Phi_p(kw,u)\,, \tag{2.45}$$

$$F_1(\nu) = \frac{H_\nu^{(1)}(kr_0)\sin(\nu[\phi_0 - \frac{\pi}{2}])}{\nu\cos(\nu\pi)}\,. \tag{2.46}$$

$$\frac{1}{2r}\int_0^{i\infty} \nu^2 \exp(i\pi\nu)\sin\pi\nu H_\nu^{(1)}(kr)A(\nu)\cos(\nu\pi)\,d\nu \\ = \frac{i}{w\cosh u}\sum_p B(\nu_p)\frac{d}{dv} f_{\nu_p}(kw,v)_{|v=3\pi/2}\Phi_p(kw,u)\,, \tag{2.47}$$

where on the joining plane we have $r = -y$, and $y = -w\sinh u$, leading to $r = w\sinh u$.

3 Spectra Derivation

In this section we derive the equations satisfied by the spectra.

Theorem 3.1 *The equations satisfied by the spectra are given by the following integral equation for $A(\cdot)$:*

$$F_1(\mu) + A(\mu)\sin(\mu\pi) = \int_0^{i\infty} K_3(\mu,\nu)A(\nu)\,d\nu\,, \tag{3.1}$$

with

$$K_3(\mu,\nu) = \frac{1}{2i}\sum_p K_1(\mu,\nu_p)\frac{1}{\frac{d}{dv} f_{\nu_p}(kw,v)|_{v=3\pi/2}} K_2(\nu,\nu_p)\,, \tag{3.2}$$

$$K_1(\mu,\nu_p) = i\, f_{\nu_p}(kw,v)_{|v=3\pi/2}\int_0^{\infty} \coth u\, H_\mu^{(1)}(kw\sinh u)\Phi_p(kw,u)\,du\,, \tag{3.3}$$

$$K_2(\nu,\nu_{p'}) = \nu^2 \exp(i\pi\nu)\sin(\pi\nu)\cos(\nu\pi)\int_0^{\infty} \coth u\, H_\nu^{(1)}(kw\sinh u)\Phi_{p'}(kw,u)\,du\,; \tag{3.4}$$

and the linear system satisfied by $B(\cdot)$

$$B(\nu_{p'}) = s(\nu_{p'}) + \sum_{\nu_p} D_{p'p}B(\nu_p)\,, \tag{3.5}$$

where

$$s(\nu_{p'}) = \frac{-1}{2i} \frac{1}{\frac{d}{d\nu} f_{\nu_{p'}}(kw,\nu)|_{\nu=3\pi/2}} \int_0^{i\infty} K_2(\nu,\nu_{p'}) \frac{F(\nu)}{\sin\nu\pi} d\nu\,, \tag{3.6}$$

$$D_{p'p} = \frac{1}{2i} \frac{1}{\frac{d}{d\nu} f_{\nu_{p'}}(kw,\nu)|_{\nu=3\pi/2}} \int_0^{i\infty} K_2(\nu,\nu_{p'}) \frac{K_1(\nu,\nu_p)}{\sin\nu\pi} d\nu\,. \tag{3.7}$$

We divide the proof into four steps. Firstly, we start by multiplying Eq. (2.45) by $\frac{1}{r} H_\mu^{(1)}(kr)$ and integrating with respect to r from 0 to ∞. To that end we make use of [8]

$$\int_0^{\infty} H_\mu^{(1)}(kr) H_\nu^{(1)}(kr) \frac{dr}{r} = D(\nu,\mu)\,, \tag{3.8}$$

$$D(\nu,\mu) = \frac{2}{\mu \sin\mu\pi} \exp\left[-i(\mu+\nu)\pi/2\right] \left[\delta(\Im m\nu - \Im m\mu) + \delta(\Im m\nu + \Im m\mu)\right]\,. \tag{3.9}$$

For the particular case which is considered here ($\Im m\mu > 0$, $\Im m\nu > 0$), the $\delta(\Im m\nu + \Im m\mu)$ term in Eq. (3.9) will be removed. Therefore, we have

$$D(\nu,\mu) = \frac{2}{\mu \sin\mu\pi} \exp\left[-i(\mu+\nu)\pi/2\right] \delta(\Im m\nu - \Im m\mu)\,. \tag{3.10}$$

Hence,

$$F_1(\mu) + A(\mu)\sin(\mu\pi) = \sum_p K_1(\mu,\nu_p) B(\nu_p)\,, \tag{3.11}$$

where

$$K_1(\mu,\nu_p) = i\, f_{\nu_p}(kw,\nu)_{|\nu=3\pi/2} \int_0^{\infty} \frac{1}{r} H_\mu^{(1)}(kr) \Phi_p(kw,u)\, dr\,, \tag{3.12}$$

and from $r = w\sinh u$, it follows

$$K_1(\mu,\nu_p) = i\, f_{\nu_p}(kw,\nu)_{|\nu=3\pi/2} \int_0^{\infty} \coth u\, H_\mu^{(1)}(kw\sinh u) \Phi_p(kw,u)\, du. \tag{3.13}$$

In a second step, multiplying both sides of Eq. (2.47) by $\Phi_{p'}(kw,u)/i\tanh u$, integrating on u from 0 to ∞, and making use of the orthonormality relation in Eq. (A.10), we obtain

$$B(\nu_{p'}) = \frac{1}{2i} \frac{1}{\frac{d}{d\nu} f_{\nu_{p'}}(kw,\nu)|_{\nu=3\pi/2}} \int_0^{i\infty} K_2(\nu,\nu_{p'}) A(\nu)\, d\nu\,, \tag{3.14}$$

$$K_2(\nu,\nu_{p'}) = \nu^2 \exp(i\pi\nu) \sin(\pi\nu) \cos(\nu\pi) \int_0^{\infty} \coth u\, H_\nu^{(1)}(kw\sinh u) \Phi_{p'}(kw,u)\, du\,. \tag{3.15}$$

In a thirst step, we substitute Eq. (3.14) into Eq. (3.11) and we obtain the integral equation satisfied by $A(\mu)$

$$F_1(\mu) + A(\mu)\sin(\mu\pi) = \int_0^{i\infty} K_3(\mu,\nu) A(\nu)\, d\nu\,, \tag{3.16}$$

$$K_3(\mu,\nu) = \frac{1}{2i} \sum_p K_1(\mu,\nu_p) \frac{1}{\frac{d}{d\nu} f_{\nu_p}(kw,\nu)|_{\nu=3\pi/2}} K_2(\nu,\nu_p)\,. \tag{3.17}$$

Finally, from Eq. (3.11) into Eq. (3.14), we obtain the linear system satisfied by $B(\nu_{p'})$:

$$B(\nu_{p'}) = s(\nu_{p'}) + \sum_{\nu_p} D_{p'p} B(\nu_p) \,, \tag{3.18}$$

where

$$s(\nu_{p'}) = \frac{-1}{2i} \frac{1}{\frac{d}{d\nu} f_{\nu_{p'}}(kw,\nu)|_{\nu=3\pi/2}} \int_0^{i\infty} K_2(\nu,\nu_{p'}) \frac{F(\nu)}{\sin\nu\pi} d\nu \,, \tag{3.19}$$

$$D_{p'p} = \frac{1}{2i} \frac{1}{\frac{d}{d\nu} f_{\nu_{p'}}(kw,\nu)|_{\nu=3\pi/2}} \int_0^{i\infty} K_2(\nu,\nu_{p'}) \frac{K_1(\nu,\nu_p)}{\sin\nu\pi} d\nu \,. \tag{3.20}$$

4 Field Representations

4.1 On the C_1 Region

The field in C_1 is given from Eq. (2.16) together with Eq. (3.18)–(3.20). From the asymptotic behavior of the radial Mathieu functions in Eq. (A.19) and that of the angular ones, the series summand is dominated by $\exp(-\gamma_p[\frac{\pi}{2}+v])$, and hence the series in Eq.(2.16) converges exponentially.

4.2 On the C_2 region

4.2.1 Near Field

In this subsection we will deduce the near field formulas.

Theorem 4.1 *We have for the $p_2^{(1)}$ contribution*

$$p_2^{(1)}(r,\phi) = -\pi i \sum_{\nu_s} \nu_s J_{\nu_s}(kr) Res[A(\nu_s)] \sin\left(\nu_s\left[\phi - \frac{\pi}{2}\right]\right) \,, \tag{4.1}$$

where the residues $Res[A(\nu_s)]$ are given by

$$Res[A(\nu_s)] = \frac{1}{\pi} \frac{H_{\nu_s}^{(1)}(kr_0)\sin(\nu_s[\phi_0 - \frac{\pi}{2}])}{\nu_s} \,, \quad \nu_s = s + \frac{1}{2} \,, \tag{4.2}$$

and

$$\begin{aligned} Res[A(\nu_s)] &= \frac{-1}{\pi} \frac{H_{\nu_s}^{(1)}(kr_0)\sin(\nu_s[\phi_0 - \frac{\pi}{2}])}{\nu_s} \\ &+ \frac{(-1)^s}{\pi} \int_0^{i\infty} K_3(\nu_s,\mu) A(\mu) d\mu \,, \quad \nu_s = s \,. \end{aligned} \tag{4.3}$$

For $r < r_0$, making use of the KL representation in Eq. (2.20), we obtain

$$p_2^{(1)}(r,\phi) = \frac{1}{2} \int_{-i\infty}^{i\infty} \nu J_\nu(kr) A(\nu) \sin\left(\nu\left[\phi - \frac{\pi}{2}\right]\right) d\nu \,. \tag{4.4}$$

From Eqs. (3.16)–(3.17), $A(\nu)$ is a meromorphic function with pole singularities located at $\nu = \pm s$, $s = 1,2,\ldots$, and $\nu = \pm(s+1/2)$, $s = 1,2,\ldots$.

Closing contours of Eq. (4.4) in the right-hand side of the complex ν-plane and collecting residue contributions, we express

$$p_2^{(1)}(r,\phi) = -\pi i \sum_{\nu_s} \nu_s J_{\nu_s}(kr) \mathrm{Res}[A(\nu_s)] \sin\left(\nu_s \left[\phi - \frac{\pi}{2}\right]\right), \tag{4.5}$$

where the residues $\mathrm{Res}[A(\nu_s)]$ are found from Eqs. (3.16)–(3.17) as:

a)

$$\mathrm{Res}[A(\nu_s)] = \frac{1}{\pi} \frac{H_{\nu_s}^{(1)}(kr_0)\sin(\nu_s[\phi_0 - \frac{\pi}{2}])}{\nu_s}, \quad \nu_s = s + \frac{1}{2}. \tag{4.6}$$

Please note that from the asymptotic behavior of Bessel and Hankel functions [3],

$$J_\nu(z) \sim \frac{1}{\sqrt{2\pi\nu}} \left(\frac{2\nu}{ez}\right)^{-\nu}, \quad \nu \to \infty, \tag{4.7}$$

$$H_\nu^{(1)}(z) \sim \frac{1}{\sqrt{2\pi\nu}} \left(\frac{2\nu}{ez}\right)^{\nu}, \quad \nu \to \infty, \tag{4.8}$$

the series summand converges as

$$\frac{1}{\nu_s}\left(\frac{r}{r_0}\right)^{\nu_s}. \tag{4.9}$$

b)

$$\begin{aligned} \mathrm{Res}[A(\nu_s)] &= \frac{-1}{\pi} \frac{H_{\nu_s}^{(1)}(kr_0)\sin(\nu_s[\phi_0 - \frac{\pi}{2}])}{\nu_s} \\ &+ \frac{(-1)^s}{\pi} \int_0^{i\infty} K_3(\nu_s,\mu) A(\mu)\, d\mu, \quad \nu_s = s. \end{aligned} \tag{4.10}$$

In this case the series summand is dominated by $\nu_s J_{\nu_s}(kr)$, leading to the behavior given by

$$\sqrt{\frac{\nu_s}{2\pi}} \left(\frac{ekr}{2\nu_s}\right)^{\nu_s}, \quad \nu_s \to \infty. \tag{4.11}$$

Thus the first few terms of the series may increase then followed by an exponential convergence of the series summand.

Finally, in order to get the total field in C_2, we only need to add the contribution from the residues of

$$p_2^{(0)}(r,\phi) = \frac{1}{2}\int_{-i\infty}^{i\infty} J_\nu(kr) \frac{H_\nu^{(1)}(kr_0)\cos(\nu[\frac{3\pi}{2} - \phi])\sin(\nu[\phi_0 - \frac{\pi}{2}])}{\cos(\nu\pi)} d\nu, \quad \phi > \phi_0,$$

$$p_2^{(0)}(r,\phi) = \frac{1}{2}\int_{-i\infty}^{i\infty} J_\nu(kr) \frac{H_\nu^{(1)}(kr_0)\cos(\nu[\frac{3\pi}{2} - \phi_0])\sin(\nu[\phi - \frac{\pi}{2}])}{\cos(\nu\pi)} d\nu, \quad \phi < \phi_0;$$

namely

$$p_2^{(0)}(r,\phi) = i\sum_{s=1}^{\infty}(-1)^s J_\nu(kr)H_\nu^{(1)}(kr_0)\cos\left(\nu\left[\frac{3\pi}{2}-\phi\right]\right)\sin\left(\nu\left[\phi_0-\frac{\pi}{2}\right]\right),$$
$$\phi > \phi_0,$$

$$p_2^{(0)}(r,\phi) = i\sum_{s=1}^{\infty}(-1)^s J_\nu(kr)H_\nu^{(1)}(kr_0)\cos\left(\nu\left[\frac{3\pi}{2}-\phi_0\right]\right)\sin\left(\nu\left[\phi-\frac{\pi}{2}\right]\right),$$
$$\phi < \phi_0,$$

where $\nu = s+1/2$.

4.2.2 Far Field

It is known [3] that the Kontorovich-Lebedev representation diverges when the rate of exponential decay of $A(\nu)$ in ν is less than the rate of exponential increase of $\{H_\nu^{(1)}(kr)\sin(\nu[\phi-\frac{\pi}{2}])\}$, and we need to continue analytically the integrand to get the far field. With the lack of an analytic formula for $A(\nu)$ such a continuation needs to be done numerically, which is very cumbersome for the problem under consideration. We propose an alternative way to calculate the far field. That is, to invoke the *reciprocity principle*. In this way, to calculate the fields when $r > r_0$, we employ

$$p_2^{(1)}(r,r_0,\phi,\phi_0) = -\pi i\sum \nu J_\nu(kr_0)\mathrm{Res}[A(\nu)]\sin\left(\nu\left[\phi_0-\frac{\pi}{2}\right]\right), \tag{4.12}$$

where the sum is on the residues of $A(\nu)$ with the source located at (r,ϕ) in C_2 and the observer located at (r_0,ϕ_0) in C_2 and we add the contribution from $p_2^{(0)}$ in the same manner as was done before.

It is worth mentioning here that the above reciprocity-based residue sum cannot be used to derive far field results for plane wave illumination since both source and observer are in the far region.

4.2.3 The Plane Wave Illumination Case

In the present subsection we obtain the field representation for the plane wave illumination case.

Theorem 4.2 *In the present case, we have the far field representation*

$$p_2^{(1)}(r,\phi) = -4i\int_\gamma \exp(ikr\cos\alpha)f_2^{(1)}(\alpha)\,d\alpha, \tag{4.13}$$

where

$$f_2^{(1)}(\alpha) = \frac{-i}{2}\sum_{\nu_s}\exp(i\alpha\nu)\exp(-i\pi\nu/2)\,\nu\,Res[A(\nu)]\sin\left(\nu\left[\phi-\frac{\pi}{2}\right]\right)_{|\nu=\nu_s} \tag{4.14}$$

with ν_s standing for the two pole sets $\nu_s = s$, $\nu_s = s+1/2$.

The far field due to a normal (with respect to z) incident unit amplitude plane wave is recovered by:

a) Letting $r_0 \to \infty$, and setting

$$\frac{1}{4}\sqrt{\frac{2}{\pi k r_0}}\exp\left[i\left(kr_0+\frac{\pi}{4}\right)\right]=1 \tag{4.15}$$

(the above, cf. [3], amounts to replacing $i[H_\nu^{(1)}(kr_0)]/4$ by $\exp(-i\pi\nu/2)$).

b) Transforming Eq. (4.4) into an equivalent one by using (see [9]):

$$J_\nu(kr)=\frac{\exp(-i\pi\nu/2)}{2\pi}\int_\gamma \exp(ikr\cos\alpha+i\alpha\nu)\,d\alpha\,, \tag{4.16}$$

where γ is the Sommerfeld integration path going from $-\pi/2+i\infty$ to $3\pi/2+i\infty$ in the complex α–plane.

We obtain the far field representation

$$p_2^{(1)}(r,\phi)=-4i\int_\gamma \exp(ikr\cos\alpha)f_2^{(1)}(\alpha)\,d\alpha\,, \tag{4.17}$$

where

$$f_2^{(1)}(\alpha)=\frac{1}{4\pi}\int_{-i\infty}^{i\infty} d\nu\exp(i\alpha\nu)\exp(-i\pi\nu/2)\nu A(\nu)\sin\left(\nu\left[\phi-\frac{\pi}{2}\right]\right). \tag{4.18}$$

By closing contours to the right-hand side of the complex ν plane and collecting residue contributions from the poles of $A(\nu)$, we obtain

$$f_2^{(1)}(\alpha)=\frac{-i}{2}\sum_{\nu_s}\exp(i\alpha\nu)\exp(-i\pi\nu/2)\nu\,\mathrm{Res}[A(\nu)]\sin\left(\nu\left[\phi-\frac{\pi}{2}\right]\right)_{|\nu=\nu_s} \tag{4.19}$$

where ν_s stands for the two pole sets $\nu_s=s$, $\nu_s=s+1/2$, as stated above. Since $\Im m\alpha>0$, one concludes by a similar way as the discussion given before that the series in the $f_2^{(1)}(\alpha)$ formula converges.

For obtaining now the total field in C_2, we just need to add the contribution

$$p_2^{(0)}(r,\phi)=-4i\int_\gamma \exp(ikr\cos\alpha)f_2^{(0)}(\alpha)\,d\alpha\,, \tag{4.20}$$

leading to

$$f_2^{(0)}(\alpha) = \frac{i}{2\pi}\sum_{\nu_s}(-1)^s\exp(i\alpha\nu)\exp(-i\pi\nu)\times \cos\left(\nu\left[\frac{3\pi}{2}-\phi\right]\right)\sin\left(\nu\left[\phi_0-\frac{\pi}{2}\right]\right)_{|\nu=\nu_s},\ \phi>\phi_0, \tag{4.21}$$

$$f_2^{(0)}(\alpha) = \frac{i}{2\pi}\sum_{\nu_s}(-1)^s\exp(i\alpha\nu)\exp(-i\pi\nu)\times \cos\left(\nu\left[\frac{3\pi}{2}-\phi_0\right]\right)\sin\left(\nu\left[\phi-\frac{\pi}{2}\right]\right)_{|\nu=\nu_s},\ \phi<\phi_0, \tag{4.22}$$

where $\nu_s = s + 1/2$, and the series converges exponentially.

Final Remark. Since the problem under consideration is one of scattering and diffraction, the wave number k is real (complex) for the lossless (lossy) case. However, for a while we have assumed that k (in the C_2 region) is such that

$$\phi = \arg k = \frac{\pi}{2} \,. \tag{4.23}$$

The condition in Eq. (4.23) has been shown by Osipov [7] to guarantee the convergence of the used KL integral representations, which in its turn is required for the possibility to use the boundary conditions conveniently, leading to the derivation of the equations satisfied by the spectra. Once those are derived, KL integral representations are obtained (cf., e.g., Eq. (4.4)). From these integral representations one derives an alternative field representation in terms of series sums (see, for instance, Eq. (4.5)). Finally, we continue the series sums analytically with respect to k. That is, extending k to real or complex values (to restore the original scattering and diffraction problem). The previous steps were not mentioned explicitly along the text only for a matter of brevity. The justification of the analytic continuation of k from imaginary to complex or real values comes from the convergence of the final series sums (see, for example, (4.9)).

5 Conclusion

The boundary-value problem for the Helmholtz equation with (prototype) Dirichlet boundary conditions in an L-shaped obstacle has been examined using Mathieu functions and the Kontorovich-Lebedev transform. The integral equation satisfied by the KL spectrum as well as the linear system satisfied by the diMf spectra have been obtained in Section 3. In Appendix A, the diMf transform has been derived. Representations for the near and far fields were given in Section 4.

The approach of this paper is applicable to problems of thermal conductivity, electromagnetics and elastodynamics with an L-shaped boundary. It is expected that additional geometries, with boundary conditions (both of the continuity and impedance types) on radial directions, could also be considered by means of the present formulation.

A Construction of the Index Transform

In order to construct the index of the radial Mathieu function transform, we make use of the method of characteristic Green's functions [3]. To that end, we start by deriving Green's function, g_u, for the modified (radial) Mathieu functions in the ODE [5]

$$g_u''(u,u_0,k^2,\lambda) - (\lambda - k^2 w^2 \cosh^2 u)\, g_u(u,u_0,k^2,\lambda) = -\delta(u-u_0)\,, \tag{A.1}$$

(where λ is a separation constant and $u_0 \in (a,\infty)$) under the conditions:

a) Boundary condition at $u = a$

$$g_u(u=a,u_0,k^2,\lambda) = 0\,; \tag{A.2}$$

b) Radiation condition as $u \longrightarrow \infty$;
c) Source conditions

$$g_u(u_0+0,u_0,k^2,\lambda) = g_u(u_0-0,u_0,k^2,\lambda)\,, \tag{A.3}$$

$$\frac{d}{du}[g_u(u_0+0,u_0,k^2,\lambda) - g_u(u_0-0,u_0,k^2,\lambda)] = -1\,. \tag{A.4}$$

Altogether, we obtain

$$\begin{aligned} g_u(u,u_0,k^2,\lambda) &= \frac{\pi i}{2}\left[J(\nu,kw,u_0) - \frac{J(\nu,kw,a)}{H^{(1)}(\nu,kw,a)}H^{(1)}(\nu,kw,u_0)\right]\times \\ & \quad H^{(1)}(\nu,kw,u)\,, \qquad u > u_0, \\ g_u(u,u_0,k^2,\lambda) &= \frac{\pi i}{2}\left[J(\nu,kw,u) - \frac{J(\nu,kw,a)}{H^{(1)}(\nu,kw,a)}H^{(1)}(\nu,kw,u)\right]\times \\ & \quad H^{(1)}(\nu,kw,u_0)\,, \qquad u < u_0, \end{aligned} \tag{A.5}$$

where $\nu = \sqrt{\lambda}$, $J(\nu,kw,u)$ (respectively $H^{(1)}(\nu,kw,u)$) are modified (radial) Mathieu functions of the first (respectively, third) type.

The 1-D Green's function, in Eq. (A.5), implies [3] the completeness relation

$$\delta(u-u_0) = \frac{1}{2\pi i}\int_c g_u(u,u_0,k^2,\lambda)\,d\lambda\,, \tag{A.6}$$

where the closed contour c is taken in the positive (counterclockwise) sense around all the singularities of g_u.

Making the substitution $\lambda = \nu^2$, one finds (see property 3. below) that the integrand in Eq. (A.6) has poles, ν_p, given by

$$H^{(1)}(\nu,kw,a)_{|\nu=\nu_p} = 0 \tag{A.7}$$

in the first and third quadrants of the complex ν–plane.

Closing the integration contour on the poles of the integrand, ν_p, in the first quadrant (owing to the Meixner condition in Eq. (2.6)), we obtain

$$\delta(u-u_0) = \sum_p \frac{-\pi i\nu_p J(\nu_p,kw,a)}{[\frac{\partial}{\partial\nu}H^{(1)}(\nu,kw,a)]_{\nu_p}} H^{(1)}(\nu_p,kw,u)H^{(1)}(\nu_p,kw,u_0)\,. \tag{A.8}$$

From Eq. (A.8), we derive the complete orthonormal set:

$$\Phi_p(kw,u) = \left(-\pi i\frac{\nu_p J(\nu_p,kw,a)}{[\frac{\partial}{\partial\nu}H^{(1)}(\nu_p,kw,a)]_{\nu_p}}\right)^{1/2} H^{(1)}(\nu_p,kw,u)\,, \tag{A.9}$$

and the orthogonality relation

$$\int_a^\infty \Phi_p(kw,u)\Phi_{p'}(kw,u)du = \delta_{pp'}\,, \tag{A.10}$$

together with the transform pair

$$f(u) = \sum_{p} F(\nu_p)\Phi_p(kw,u)\,, \tag{A.11}$$

$$F(\nu_p) = \int_a^\infty f(u)\Phi_p(kw,u)du\,. \tag{A.12}$$

It worth mentioning that a similar index transform, on Hankel functions, has been derived and used before in [3] to analyze diffraction by an impedance right circular cylinder in an infinite medium.

The following properties of radial Mathieu functions are also useful for our analysis:

1. Since

$$H^{(1),(2)}(\nu,kw,u) = H_\gamma^{(1),(2)}(kw\cosh u)[1+O(1/\cosh u)]\,, \quad u\to\infty\,, \quad \nu \text{ fixed}\,, \tag{A.13}$$

where $H_\gamma^{(1),(2)}(z)$ is the Hankel function of index γ of the first (second) type and $\gamma=\gamma(\nu,kw)$ is the characteristic exponent of Mathieu functions, from (cf. [5])

$$H_\gamma^{(1),(2)}(z) \sim \sqrt{\frac{2}{\pi z}}\exp\left(\pm i[z-\gamma\pi/2-\pi/4]\right)\,, \quad z\to\infty\,, \quad \gamma \text{ fixed}\,, \tag{A.14}$$

we obtain

$$H^{(1),(2)}(\nu,kw,u) \sim \sqrt{\frac{2}{\pi kw\cosh u}}\exp\left(\pm i[kw\cosh u-\gamma\pi/2-\pi/4]\right)\,, \quad u\to\infty\,, \quad \nu \text{ fixed}. \tag{A.15}$$

2. For large ν

$$\nu^2 = \gamma^2 + O(k^4w^4\gamma^{-2})\,. \tag{A.16}$$

3. From Eqs (A.13) and (A.16), the distribution of $\{\nu_p\}$ in the complex ν–plane resembles that of the zeros $\{\gamma_p\}$ of

$$H_{\gamma_p}^{(1)}(kw\ \cosh a) = 0\,, \tag{A.17}$$

namely that they exist in the first and third quadrants of the complex ν–plane (cf. [3]), symmetric around the origin and having for the first quadrant poles

$$\Im m\nu_p > \Im m\nu_q \quad \text{for} \quad p>q\,. \tag{A.18}$$

4. From [5, Eq. (20.4.8)] and the behavior of Hankel functions associated with $|\gamma_p|\to\infty$ [3, Eq. (6.7.16c)], we obtain

$$H^{(1)}(\nu_p,kw,u) = O\left(\frac{1}{\sqrt{\eta}}\exp\left[\frac{\eta\pi}{2}\frac{\ln(\cosh u/\cosh a)}{\ln(2\eta/ekw\cosh a)}\right]\right)\,, \quad \eta=|\gamma_p|\to\infty\,, \quad u \text{ fixed}\,. \tag{A.19}$$

References

[1] L. P. Castro, F.-O. Speck, and F. S. Teixeira, Explicit solution of a Dirichlet-Neumann wedge diffraction problem with a strip, *J. Integral Equations Appl.*, **15** (2003), pp. 359–383.

[2] A. Böttcher, Yu. I. Karlovich, and I. M. Spitkovsky, *Convolution Operators and Factorization of Almost Periodic Matrix Functions*, Birkhäuser, Basel, 2002.

[3] L. B. Felsen and N. Marcuvitz, *Radiation and Scattering of Waves*, Prentice-Hall inc., New Jersey, 1973.

[4] D. S. Jones, *The Theory of Electromagnetism*, Pergamon Press, New York, 1964.

[5] M. Abramowitz and I. Stegun, *Handbook of Mathematical Functions with Formulas, Graphs, and Mathematical Tables*, Dover Publications inc., New York, 1992.

[6] M. L. Kontorovich and N. N. Lebedev, On a method of solution of some problems of diffraction theory and relative problems, *J. Experim. and Theor. Physics*, **8** (1938), pp. 1192–1206.

[7] A. V. Osipov, On the method of Kontorovich-Lebedev integrals for the problems of diffraction in sectorial media, Problems of Diffraction and Propagation of Waves, *S. Petersburg University Publ.*, **25** (1993), pp. 173–219.

[8] G. Z. Forristall and J. D. Ingram, Evaluation of distributions useful in Kontorovich-Lebedev transform theory, *SIAM J. Math. Anal.*, **3** (1972), pp. 561–566.

[9] I. S. Gradshteyn and I. M. Ryzhik, *Table of Integrals, Series, and Products*, Sixth ed., Academic Press, San Diego, 2000.

INDEX

A

B

C

D

E

F

G

H

I

J

K

L

M

N

O

P

W

Y